SOVIET SUBMARINES

Above: *The 'Delta III' has become the mainstay of the Soviet SSBN forces both in the Northern and the Pacific Fleet areas. Note the anechoic tiles missing from the stern section of this Northern Fleet unit. (MoD)*

SOVIET SUBMARINES

1945 to the present

John Jordan

ARMS AND
ARMOUR

Contents

First published in Great Britain in 1989 by Arms & Armour Press, Artillery House, Artillery Row, London SW1P 1RT.

Distributed in the USA by Sterling Publishing Co. Inc., 387 Park Avenue South, New York, NY 10016-8810.

Distributed in Australia by Capricorn Link (Australia) Pty. Ltd, P.O. Box 665, Lane Cove, New South Wales 2066, Australia.

British Library Cataloguing in Publication Data
Jordan, John, *1947–*
Soviet submarines.
1. Union of Soviet Socialist Republics, Voennomorskoi flot. Submarines, history
I. Title
623.8′257′0947
ISBN 0-85368-941-5

Designed and edited by DAG Publications Ltd. Designed by David Gibbons; edited by Michael Boxall; layout by Anthony A. Evans; typeset by Typesetters (Birmingham) Ltd, Smethwick, West Midlands; camerawork by M&E Reproductions, North Fambridge, Essex; printed and bound in Great Britain by The Bath Press, Avon.

Right: *An 'Echo II' with its distinctive Snoop Slab air surveillance radar in the raised position. The submarine has a damaged bow. (MoD)*

Preface

This book represents many years of research. The study of Soviet submarines remains something of 'black art', a subject shrouded in official secrecy on both sides of the Iron Curtain. The author's aim was not therefore to produce a definitive reference work, but to further our understanding of the processes – political, philosophical and technological – which underlie the development of the Soviet submarine force during the post-war period.

The major post-war submarine types are given full coverage; this includes an analysis of design and weapons philosophy, detailed line-drawings and data tables for the various sub-groups, and a selection of photographs, many of which have not previously been published. Missile trials and other auxiliary submarines are considered peripheral to this study; they are dealt with in context, wherever it is thought to be relevant to the account of Soviet submarine development.

All of the standard reference works were consulted in compiling the data tables, but the reader needs to bear in mind the constraints under which so much of this data is gathered. Published figures are often incomplete, inconsistent, and even contradictory. The author has therefore subjected much of this data to critical analysis in an attempt to arrive at a more consistent, more coherent view. Major departures from established sources are fully documented in the text.

The drawings are based on a study of many photographs. Given the small number of open-source photographs of some types of Soviet submarine, particularly the most recent classes, this is a process fraught with problems. The drawings showing the mast-mounted sensors had to be compiled using a variety of photographs taken from different angles, as submarines are rarely photographed with all masts in the raised position. Some drawings must therefore be regarded as provisional. Nevertheless, the author is confident that this is the most complete, and the most accurate series of drawings of Soviet submarines yet published. The drawings are used both to complement the text and to help the reader in the interpretation of photographs.

A number of people have assisted me in the compilation of this book. I should particularly like to thank Antony Preston of *Defence* magazine for his help with photographic illustration; also Przemysław Budzbon, who provided many of the photographs of the warbuilt and early post-war submarines. The considerable assistance of the News Photo Branch of the Office of Information, US Navy, is also acknowledged. Finally, I should like to thank my wife, Susan, for the patience and understanding she has shown throughout this time-consuming project.

John Jordan
December, 1988

Introduction

In recent years we have become accustomed to receiving relatively accurate, and often detailed, Western Intelligence analysis of the characteristics of Soviet naval vessels while they are still under construction. Aerial surveillance of Soviet shipyards is now capable of penetrating the blanket of secrecy beneath which the Soviets habitually conceal all military activity to such an extent that the US Department of Defense is now able to publish artists' impressions of the new ships before they have even put to sea.

Inevitably, given the almost toal absence of comment on the technical aspects of their warships in Soviet written sources, characteristics still have to be deduced largely from detailed photographic analysis. Photographs of surface ships can yield a considerable amount of information. Even the approximate frequency of radar systems can be deduced by measuring the size of the reflector. Comparisons with similar Western systems can also be a fruitful source of knowledge regarding the purpose and performance of the various items of equipment aboard. Reasonably accurate assessments of the size, type and power of the propulsion plant can be made by careful study of the configuration of the exhaust uptakes and of the ventilation/air intake ducting.

The major problem in attempting to analyse the characteristics of Soviet submarines using the same procedures is that a much greater proportion of the equipment is concealed within the hull. Periscopes, radar, electronic support measures (ESM), and communications masts can be identified when they are in the raised position, as is generally the case when the boat is running on the surface. When confronted with prominent hatches or protrusions in the casing of the hull or the fin, however, we can only make assumptions about the precise nature of the equipment they conceal.

Data concerning submarine armament and propulsion systems published in the major Western reference sources are inherently unreliable and frequently contradictory. Figures have been subject to constant revision as new evidence, emanating either from photographic Intelligence or from Soviet professional writings, has come to light. For some twenty years, Soviet submarines of the 'Hotel'/'Echo'/'November' ('HEN') generation were credited with only a single nuclear reactor of 22,500–30,000shp, but, in the early 1980s – by which time some of the earlier boats were already being discarded – this assessment was revised, and they are now generally credited with two reactors each of 15,000shp.

There remain serious disagreements between British and US sources regarding the installed horsepower of some of the latest Soviet submarine types, with US sources giving consistently higher figures. The extent to which some of these figures are influenced by considerations other than accurate assessment must be open to question. High performance figures ascribed to the submarines of a potential enemy provide an excellent lever for obtaining funding not only for high-performance submarines of one's own, but also for costly anti-submarine warfare programmes. Some of the figures emanating from Western Intelligence sources may therefore be for domestic political consumption and cannot be accepted uncritically.

Some philosophical problems

The problems involved in visual analysis pale into insignificance when compared with those of what I shall call 'philosophical' analysis. Clearly, even an assessment of the capabilities of Soviet submarines by visual methods requires a degree of *interpretation*. An assessment of a mysterious tail-fin pod as an advanced propulsion system involves certain assumptions regarding Soviet technological capabilities and operational philosophy. It assumes that Soviet military scientists have been capable of bringing such a developmental concept to fruition, and that the Soviet military/industrial complex has been able to put it into full-scale production, given that manufacture may involve sophisticated computer-aided design and production processes. Finally, and most importantly, it assumes that an advanced 'quiet' propulsion system was considered essential to enable the submarine to perform its designed mission at the time when the type was first projected.

In this context it is necessary to point out fundamental differences in the way military hardware is developed in the West and in the Soviet Union. In the West, staff requirements for particular items of equipment are generally based on technology that is already available or which will be available in the immediate future. The impetus for change in the West frequently comes from the private sector, which is motivated by the need to compete for government contracts. New concepts frequently become possible because of technological developments taking place outside the framework of government-directed research and are subsequently adopted by the military.

In the Soviet Union, on the other hand, technological development is driven purely by military requirements. Those requirements are decided not at the level of a single service, as is frequently the case in the West, but at the highest political level. Overall military policy has to be determined within the framework of Marxist-Leninist philosophy. It is not sufficient for that policy to be 'effective'; it also has to be 'correct'. In the West, warfare is often talked of as 'an art'; in the Soviet Union, it is 'a science'. Technological development is therefore not something that occurs piecemeal or haphazardly; neither is it an 'opportunity' for new and more powerful weapons systems, as it so frequently is in the West. It is planned, and has to be justified in theoretical terms before it can be embarked upon. Military requirements, established within a strict framework of Marxist-Leninist dialectic, precede research and development.

Several important considerations relevant to this study follow on from these conclusions. The first is that Soviet submarine development cannot be expected to follow the same pattern as post-war developments in the West. Soviet submarines are designed to perform missions dictated by Soviet military strategy using Soviet military tactics. They may therefore be very different in overall size and configuration, and may employ different weapons systems. There may be important differences of emphasis between the various elements that make up a submarine's 'performance': speed, manoeuvrability, endurance, 'quietness', sensor/weapons capabilities, communications.

The second is that we must expect some Soviet developments to be *innovative*, without a parallel in the West. Soviet military policy, although focused against specific Western 'threats', is independent of Western thinking and is determined within its own strict ideological framework. It therefore follows that Soviet solutions will be adopted to specifically Soviet problems.

If we insist on using Western developments as a yardstick by which to measure Soviet capabilities and technological prowess, we fall into the trap of restricting our assessment to terms such as 'superiority' or 'backwardness', when what we should really be considering is why advances in a particular area of

Left: *A number of theories have been advanced regarding the precise nature of the distinctive pod mounted atop the vertical rudder-fin of recent Soviet nuclear-powered attack boats. The pod has been variously described as a housing for a towed linear sonar array, an advanced magneto-hydrodynamic 'quiet' propulsion system, and a dispenser for a towed torpedo decoy. Each of these theories involves assumptions not only about Soviet technological capabilities, but also about the operational deployment of the submarines. (Royal Navy)*

submarine technology are so important to the Soviets, and why certain other areas have been neglected.

The Soviet predilection for military innovation has been well documented. It derives in part from a faith – often misplaced – in the ability of science to resolve all problems, both human and technical. The Soviets believe that provided the initial *concept* is right, science will eventually resolve all the difficulties inherent in its realization; it being simply a matter of perseverance and time. Whereas Western military science tends towards evolutionary development, and is frequently deterred from further development in a particular direction by initial failures, Soviet developments are frequently revolutionary, and the Soviets can be remarkably persistent in the face of early disaster in their attempt to bring an important concept to fruition.

We must therefore guard against greeting innovative Soviet military systems with uncritical acclaim, and viewing them as evidence of technological superiority. In the past, some of these innovations have been unqualified disasters, to the extent that ships and submarines have been subsequently regarded by the Soviet Navy as unfit for combat. Yet because of Soviet faith in the ability of science to resolve all problems given time, these types have often by-passed the prototype stage and entered series production before the serious nature of their defects have become apparent.

The counter-side of Soviet innovation is Soviet 'backwardness'. Again, this is a word which needs to be treated with caution. Because it is driven by policy rather than by commercial or other extraneous pressures, Soviet military technology does not advance on a broad front, but is channelled into relatively narrow, predetermined areas. This means that other areas that are not central to Soviet missions can remain almost totally neglected. The classic naval example in this respect is the lack of attention given to anti-submarine warfare by the Soviet Navy during the 1950s. Since the threat from Western submarines during this time was considered negligible, there was little point in allocating vast resources to the development of sonars or anti-submarine weapons systems. When the nature of the threat changed with the advent of the Polaris submarine, a major anti-submarine programme had to be instituted, and advanced systems were rushed into development. Even so, it took the best part of ten years before these systems began to enter service in any numbers.

In contrast, when Western countries are 'caught out' by a new threat, the technology necessary to counter it is generally on hand and readily available. When deficiencies in the anti-missile capabilities of Royal Navy warships deployed in the Falklands Conflict of 1982 were discovered, ships preparing to deploy were fitted with commercially developed equipment purchased 'off-the-shelf'. The US Navy, which in late 1987 found itself in similar difficulties when confronted by a mine threat for which it was inadequately prepared, would have had no difficulty in purchasing the most advanced minehunting and minesweeping equipment from private industry had it been able to provide a suitable hull in which to install it.

It should therefore not be assumed that Soviet 'backwardness' in some areas is due to inadequacies in Soviet technology. Often it is the result of policy decisions that have determined the equipment needed to perform Soviet missions. The notorious noisiness of early Soviet nuclear submarines does not necessarily indicate that the Soviets were incapable of making them quieter. It may simply mean that noise level was

Right: *The 'Alfa' is generally regarded as the most innovative of contemporary Soviet submarine designs. With its titanium hull and 'dense' nuclear propulsion plant, the 'Alfa' is reportedly capable of operating at depths of up to 700m, and can achieve underwater speeds in excess of 40 knots. However, available evidence indicates that the design has not proved as successful as many Western commentators have suggested. The prototype was dismantled four years after entering service, and the production series was halted after only six units had been completed. (US Navy)*

Above: *The 'Victor', a near-contemporary of the 'Alfa', is a solid mainstream design in the classic Soviet mould. While the 'Alfa' has attracted the attention of the Western defence media, the 'Victor' has been entering service at the rate of a steady two/three boats per year. At the time of writing, no fewer than 45 units have been completed. Each new sub-group has introduced incremental improvements. (MoD)*

not a major consideration when these submarines were designed. Silencing a submarine is a particularly costly process; it implies a larger and more complex submarine for a propulsion plant of a given output. If silent operation is not essential to the effective performance of the submarine's mission there is no real incentive to undertake such measures, particularly when they may detract from other aspects of submarine performance, such as high underwater speed.

It would be a mistake to ascribe the advances in 'quieting' noted in the latest Soviet submarine types exclusively to espionage, or to dubious commercial practice by some Western industrial concerns. There is considerable evidence to suggest that Soviet interest in silencing techniques pre-dates by some years the technology transfer in question. The significant increase in beam which is a feature of the nuclear-powered attack submarines of the 'Sierra' and 'Akula' classes is probably indicative of the adoption of a raft-mounted machinery installation, which was incorporated into Western SSNs from the early 1960s onwards, and which could equally well have been incorporated into Soviet submarines of an earlier generation had it been considered desirable. This suggests that the impetus for Soviet activities directed at acquiring Western technology was provided by an important change in Soviet tactical/operational policy regarding the employment of their attack submarines.

Soviet advances in silencing therefore have less to do with 'technological opportunism' than with a policy change requiring rapid technological progress in an area previously neglected by Soviet military science. There can be little doubt that the technology transfer that has occurred has saved the Soviet Union a considerable amount of time and effort which otherwise would have had to be expended on research and development. However, the technology has been acquired not for its own sake, or because the West was seen to be 'ahead' in this particular area, but because it was required in order to allow the Soviets to fulfil their own policy objectives.

Soviet naval policy

Since an understanding of Soviet policy objectives is fundamental to any study of Soviet submarine development, it is worth looking more closely at how naval policy is formed.

Soviet military strategy is determined at the highest political level, and has to be based on Marxist-Leninist theoretical teachings about war and the army. The Defence Council (*Sovyet Oborny*), which is chaired by the General Secretary of the Communist Party, is made up of selected members of the Politburo, and the heads of the Ministry of Defence, the Council of Ministers, and GOSPLAN (the state economic planning agency), together with the Chief of the General Staff. The Defence Council not only establishes overall military policy goals, but determines the means by which these goals are to be achieved. It has full control over the military budget, and decides which developmental projects should be pursued, and the number and balance of major weapons system purchases.

The senior peacetime body of the armed services is the Main Military Council (*Glany Voyennikh Sovyet*), which advises the Defence Council on strategy and operations. The Main Military Council is chaired by the Minister of Defence, and comprises the four First Deputy Ministers (including the Chief of the General Staff), the five military heads of the armed services, and six other senior defence officials at the head of the various military agencies.

In theory the degree of influence that any single service can wield is extremely small. In practice the Soviet Army provides all but a handful of the members of the Main Military Council (including all six Heads of Agencies). The Soviet Navy, in contrast, occupies only one seat on the council.

There can be little doubt that some Soviet senior naval officers have in the past been less than happy with this situation. Arguably, the primary motivation for Admiral Gorshkov's influential work *The Sea Power of the State* (which first appeared in 1976 and was subsequently translated into English),[1] was an attempt to establish the concept of 'sea power' as an independent area of study. Underlying the historical analysis, which is deemed necessary in Soviet military writings in order to support the 'objectivity' of one's arguments, is a plea for naval solutions to specifically naval problems.

This view runs counter to traditional Soviet military thinking, and it has been opposed strongly not only outside the Navy, but by influential figures within the Navy itself. When Vice-Admiral Stalbo, who has been long associated with Gorshkov, advanced similar views in a major two-part article published in 1981,[2] he was subjected to an immediate verbal assault from all sides. Many of his critics were political officers holding prominent posts in the Soviet naval academies. They argued strongly against Stalbo's 'narrow view' that naval theory should be treated as if it were independent from the study of military science, and stated firmly that naval problems should be examined at a higher level than the Soviet Navy.[3]

Of even greater significance, however, were the attacks emanating from prominent serving naval officers, most notably Admiral Chernavin (then Commander-in-Chief, Northern Fleet) and Admiral Kapitanets of the Baltic Fleet. Chernavin's criticisms are particularly pertinent in view of his subsequent accession to the post of Commander-in-Chief of the Soviet Navy. He accused Stalbo of 'subjective' interpretation and of a 'lack of unified views', and reminded him that Marxist-Leninist philosophy recognizes 'immutable' laws of society and war, rigid methods of analysis of given military situations, and ideologically correct solutions.

It is clear, therefore, that we are not talking about an 'inter-service' battle such as frequently occurs in the West, but about a far more fundamental debate about naval policy which has been taking place within the Soviet Navy itself. Whereas inter-service battles in the West are generally about budgets and equipment, the debate in the Soviet Union has a philosophical dimension. It is about who is best qualified to determine the ways and means by which state military policy as it affects the maritime sphere of operations should be implemented. It is also about whether naval operations constitute a separate branch of military science with its own disciplines, and its own special problems susceptible to its own particular solutions.

The vigorous and often heated debate on the US Navy's Maritime Strategy which has been going on over the past few years in the United States puts an interesting perspective on the Soviet debate. It could be argued that the new US maritime policy has been arrived at by a similar process to that advocated by Gorshkov, in that it was drawn up in response to the Soviet maritime threat, and proposes naval solutions to naval problems. Its major critics outside the US Navy have focused on the distortion of the overall strategic picture which they claim is inherent in the Maritime Strategy, and in particular the failure to relate its projected sequence of naval operations to what would be happening on land. They have claimed that the Maritime Strategy is a recipe for success in the peripheral theatres of war at the expense of total and disastrous failure in the central (continental) theatre. Some critics have suggested that there is little point in having a policy based on prolonged conventional operations by strong naval forces in the maritime theatre when the NATO land forces are so feeble that they would quickly have to resort to the first use of tactical nuclear weapons in order to contain the Soviet armoured divisions.

The first point of interest is that the concept of a 'Maritime Strategy' is not considered unusual or unnatural in the West. The second is that the criticisms outlined above, although couched in very different language, are in effect identical with those advanced in the Soviet Union against the views of Gorshkov and Stalbo. Essentially the US Navy's policy document is being criticized as 'subjective' (i.e., promoting single-service interests by focusing on naval priorities at the expense of an 'objective', overall view of the state's defence interests), and for a 'lack of unified views'. The US Maritime Strategy debate casts an interesting light on the very different way in which military policy is

arrived at in the West, but it also reveals that the criteria against which the merit of such strategies must be judged are essentially the same.

The military-industrial economy

A further important constraint on Soviet submarine development has been the nature of the Soviet military-industrial complex. The influence of GOSPLAN, the state economic planning agency, on Soviet military programmes should not be underestimated. As was noted earlier, GOSPLAN is represented on the Defence Council (i.e., at a higher political level than the Soviet Navy). It is therefore influential in deciding what should and what can be done. There is no question of the military deciding what it wants and then putting the business out to tender, as is generally the case in the West. Because GOSPLAN is influential at the level of policy-making, the design process of Soviet military systems has to take the production process into account.

Soviet industry's capacity for high technology production continues to be relatively small. Some military projects (notably the Soviet space programme and the 'Alfa'-class submarine) reveal high-technology capabilities on a par with, or even in advance of, the West. These programmes, however, should be seen as the apex of a triangle with an exceptionally broad, low-technology base. At the lowest level, Soviet industry appears to find it difficult to produce a reliable television set for popular consumption.

Most Soviet military systems are designed for ease of operation, ease of maintenance, and ease of production. The first two considerations relate to the high proportion of short-term conscripts in the Soviet armed forces; the third relates to the need to produce large quantities of equipment in a variety of relatively low-technology factories. The construction of six 'Alfa'-class submarines is undoubtedly a remarkable achievement by Soviet industry, but it is an achievement that needs to be put in its proper perspective. It would have been much more surprising if the advanced technology related to the 'Alfa' had been incorporated into the contemporary SSNs of the 'Victor' class, which numbers 46 units. The Soviet military industry is capable of creating an 'Alfa', but is less likely to produce an M1 Abrams main battle tank.

The continuing production of diesel-powered submarines to serve alongside nuclear boats is less a matter of choice than of necessity. The Soviet Union is incapable of building the number of nuclear boats it would ideally like. Indeed it was only possible to change the balance of construction in favour of nuclear-powered submarines from the mid-1960s following a policy decision to equip three further shipyards (in addition to Severodvinsk and Komsomolsk) for nuclear construction. Manpower considerations have also been influential, as nuclear machinery is inherently more complex to operate and to maintain. Soviet nuclear submarines spend a considerable part of their lives tied up inside Soviet bases, constituting a 'ready reserve' to be deployed in the event of large-scale mobilization.

The 'high-low' mix evident in Soviet naval construction is frequently also evident within the high-technology programmes themselves. In any given system, only those features essential to high performance will incorporate advanced technology. Other sub-systems may well employ equipment already in production with a much lower technology rating.

Above: *The primary mission of the submarines built for the Soviet Navy prior to the Second World War was the defence of Soviet waters against incursions by hostile surface vessels. The unit depicted is an 'L'class 'minelayer', a type developed from the British submarine* L55, *which was sunk in the Baltic in 1919. (Przemyslaw Budzbon collection)*

There is an inherent inertia in Soviet military production which can result in delays and 'bottlenecks' in the supply of new items of equipment. The dominance of the 'Five Year Plan', with its emphasis on production targets, encourages long runs and provides factory managers with a powerful disincentive to disrupt production lines for retooling. It is quicker to modify existing equipment than to redesign it. Hence the Soviet predilection for add-on modifications, often at the expense of other aspects of performance. Where possible, new designs incorporate as much existing technology as possible.

Modification is also preferred to redesign when it comes to the total weapons system. The 'Victor'-class submarine has been in production for more than twenty years, and has been subjected to two major modifications. The basic design of the ballistic missile submarines of the 'Delta IV' class, which have only recently entered production, dates from the early 1960s, being derived from the 'Yankee' via three earlier variants of the 'Delta'. The employment of what we in the West would consider to be obsolescent technology is of little concern to the Soviets, provided that it does not detract from the primary mission of the submarine. In the case of the 'Delta IV', the hull and machinery design is clearly considered adequate to support deployment of the high-technology SS-N-23 ballistic missile, which constitutes the main armament of these SSBNs.

Below: *The recourse to 'mirror-imaging' by Western commentators when attempting to analyse Soviet intentions and capabilities leads to frequent errors of interpretation. Soviet operational/tactical thinking, which is dominated by the Combined Arms concept, is very different from that which prevails in Western navies, where the submarine still tends to be regarded as a 'lone wolf'. Here we see Lieutenant-Captain V. Motrenko, described in the caption as the 'navigator of a submarine formation' of the Northern Fleet. (TASS)*

Strategy and tactics

There are fundamental differences between Soviet and Western concepts of seapower. Western strategists tend to view the sea in a positive way. The sea is important for communication and for trade; it can also be used to bring military pressure – either by blockade or by direct assault – on an adversary. Western maritime theory as defined by the pre-eminent American strategist, Alfred Thayer Mahan (*The Influence of Seapower upon History, 1660–1783*), is therefore based on the twin pillars of 'sea control' and 'power projection'.

The Soviet view of the sea is essentially negative. For the Soviets the sea represents not an 'opportunity' but a vulnerable flank. The concept of overseas trade and communication so crucial to the West is, for the Soviets, inextricably linked with colonialism and the capitalist economy. That is not to say that seaborne trade and communications have no place in Soviet thinking. On the contrary, the experience of the past 25 years suggests that the Soviets have developed an increasing awareness of the importance of such trade links. The Soviet merchant marine has made significant inroads into the traditional shipping routes and has earned valuable foreign currency in the process. It has also been extensively employed to provide both political and material support to friendly, developing countries.

However, it is important to see these as essentially peacetime operations. The Soviet Union is not dependent on the ocean trade routes either for raw materials and foodstuffs, or for the transport of troops and *matériel.* Consequently the only Soviet sea lines of communication that would require protection in the event of war would be those sea routes close to the Soviet littoral used by the Soviet Union for internal communication.

As a continental land power, the Soviet Union sees itself as a potential victim of the 'power projection' philosophy espoused by the major Western seapowers, and Soviet naval strategy has traditionally focused on defeating Western seaborne assault. Defence of the homeland therefore continues to be a primary mission of the Soviet Navy.

Soviet military theory determines not only Soviet missions, but the ways and means by which military objectives are to be attained. Soviet military tactics are considered too important to be left in the hands of the commander in the field. They are essentially *collaborative*, and are subject to a unified command structure which parallels the policy-making apparatus. The initiative of the individual, which has so often made the difference between success and failure in the great sea battles of the past, has no place in the Soviet concept of naval operations. To depend on the character and judgement of any one individual for the success of a major military operation would be 'unscientific'. All participants in a military operation, including the commanders, are therefore expected to follow a 'correct' and closely-defined course of action. These tactics are essentially similar to those employed by the Soviet ground forces, from which they are clearly derived.

In the West we see warfare on land as a totally different discipline. Army Fronts are continuous and in constant contact with the enemy, and the divisions of which they are composed are interdependent. There would be no question of moving one division forward without using other units to secure the flanks. Modern land operations are also multi-dimensional, with aircraft providing essential reconnaissance behind the enemy front, together with tactical support for friendly ground forces. The generals control and co-ordinate operations from secure HQs behind the fronts, and their decisions are based on situation maps which give them a complete picture of operations in their area.

Open ocean operations, on the other hand, have traditionally been characterized by the physical isolation of individual units and task forces, fleeting contact with the enemy, and the difficulty of providing sustained air support from continental airfields. Sea command is characterized by the need to make rapid decisions, often on the basis of incomplete or inaccurate information.

Western naval forces have evolved as essentially self-contained task groups, capable of performing their predetermined missions independently. They provide reconnaissance within their own immediate area of operations, and can defend themselves against attack from any quarter, whether it be on, above or below the surface. Western tactical theory has traditionally focused on bringing the enemy's battlefleet to action and defeating it, thereby securing control of the sea.

The 'Young School'

In the early years following the Revolution, there was no question of the Soviet Navy building a battlefleet capable of challenging any of the major sea powers. The Soviet Union had neither the resources nor the shipbuilding capability to construct such a fleet. An alternative strategy was therefore required.

The so-called 'Young School' of naval officers which emerged from the naval academies during this period proposed a new naval strategy consistent with revolutionary ideals and Soviet security needs. Central to this strategy was the idea of the sea as a potential 'buffer' against assault by hostile powers on the periphery of the Soviet land mass. The sea was no longer to be regarded as an operational area independent of land, but was to be integrated into the various land-based theatres of operations.

Naval tactics would henceforth be an extension of those developed for the Soviet Army. They would stress co-operation between the various forces available, together with rigid central control from command posts based not aboard a flagship but on land. The 'Combined Arms' philosophy, as it became known, was made possible by the proximity of the Navy's major operational areas (initially in the Baltic and Black Seas) to the Soviet land mass. The new technology of the 1920s and 1930s held out the promise of a powerful force of land-based bombers capable of striking at any hostile battlefleet that dared to enter Soviet sea-space, and of long-range fighters that could provide a protective 'umbrella' above the Soviets' own surface forces.

The other military system that appeared to be particularly well-suited to Soviet missions of the period was the submarine. The submarine was the ideal weapon for the defence of Soviet sea-space. Because of the overall dominance of the enemy battlefleets, Soviet surface forces would generally have to be deployed as 'interceptors', and would require accurate aerial reconnaissance if they were to perform their missions satisfactorily. Submarines, on the other hand, could be deployed in patrol lines across the path of advance of an enemy battlefleet. They would then be ideally placed to make their attacks from the port or starboard bow of the enemy line.

Submarines could also be used offensively without the need for sea control. Unlike surface forces they could operate in sea areas where the surface was controlled by the enemy. They could attack merchant shipping, lay mines in the approaches to ports and harbours, and provide information on the movements of hostile naval forces.

The Soviet Navy was quick to see the potential of the submarine, and made strenuous efforts to acquire the necessary technology from abroad. Virtually all the early Soviet medium- and long-range submarines were based on foreign designs. The 25 'L'-class boats of the 1928–32 programme were modelled on the British 'L' class of the First World War, while many of the submarines built under later programmes were based on German designs.

Time, axis, location

Because sea control is not essential for the performance of Soviet missions, the Soviet Navy does not need to maintain naval forces superior in number or in fighting power to those of its enemies. The tables showing the number of vessels in each of the different categories operated by the respective world navies that are an essential feature of all Western reference books have no place in Soviet thinking. For the Soviets they represent a hangover from the days of the battlefleets of a previous era, when totting-up the number of dreadnoughts and super-dreadnoughts was sufficient to reveal which navy ruled the waves.

The concept of the 'fleet action', in which the major part of the enemy's surface forces is brought to battle by the main body of one's own surface forces in order to determine mastery of the seas continues to play an important part in Western naval thinking. The history of two world wars teaches us that fleet actions such as the Battle of Midway, and even the relatively inconclusive Battle of Jutland, have often been decisive in determining the outcome of a much wider conflict. The fleet action is also crucial to the US Navy's Maritime Strategy, in that the primary reason for thrusting the carrier striking force into the northern reaches of the Norwegian Sea is to compel the Soviet Navy to come out and fight on terms of the US Navy's choosing.

However, Soviet naval strategy has been evolved specifically to counter this type of 'forcing' action. The Soviets have long recognized that they cannot engage the major Western battlefleets on equal terms. Soviet tactical thinking is therefore based on the sort of 'ambush' which, in the context of land operations, one might direct against an armoured column. The overall correlation of strength between the two sides is relatively unimportant in this particular equation. The crucial factor is the achievement of local superiority at a time and place of the defenders' choosing, preferably with an element of surprise. This style of tactical thinking is generally attributed to Rear-Admiral V. A. Belli, whose theories of naval engagement have frequently been quoted by senior naval officers such as Gorshkov.

Local superiority is to be achieved by gathering all the forces (surface ships, aircraft and submarines) available in the relevant theatre of operations, and by mounting a centrally controlled 'Combined Arms' assault on the enemy fleet at a predetermined location. Near-simultaneous strikes would then be mounted from a variety of axes in order to disrupt the enemy formation and overwhelm the defences. The prerequisites for such tactics are accurate reconnaissance, secure (and preferably real-time) communications, and a high level of training.

Science and history

It has been noted earlier in this chapter that the Soviets believe that history is on their side and that science, given time, will provide solutions for even the most difficult technical problems.

There can be little doubt that modern technology has made 'Combined Arms' operations more feasible, and potentially more effective. Broad-area ocean reconnaissance systems based on satellites are now employed extensively, both by the West and by the Soviet Navy, to monitor the movement and electronic emissions of surface ships. There has likewise been a marked improvement in communication between ships, between ship and aircraft, and between ship and shore.

Below: *Co-operation between submarines and surface ships is a regular feature of Soviet deployments and exercises. Here a diesel submarine of the 'Foxtrot' class is seen in company with a 'Kashin' class destroyer in Mediterranean waters.*

Above: *Many modern Soviet attack submarines carry long-range cruise missiles to counter Western 'power projection' forces such as carrier battle groups. Seen here is a unit of the 'Echo II' class, armed with eight 200nm SS-N-3 Shaddock missiles. (MoD)*

The Soviet Navy is not alone in having benefited from these developments. Western anti-submarine warfare tactics have become increasingly dependent on similar underwater broad-area ocean surveillance systems such as SOSUS for their effectiveness, and 'Combined Arms' forces comprising surface ships, submarines and land-based 'pouncer' aircraft such as the Nimrod and P-3 Orion are deployed under centralized land-based control within a command structure not dissimilar to that favoured by the Soviets.

Weapons technology has also moved on, and recent developments have been particularly favourable to the Soviet 'ambush' concept. Surface ships and aircraft no longer have to approach to within visual range in order to engage the enemy, but can release their anti-ship missiles from a distance of 200–300 nautical miles. This facilitates the launching of near-simultaneous saturation strikes, and makes the launch platform more difficult to locate and destroy.

The increased range of modern weapons has its counter-side, however. Although Soviet ships and aircraft are now capable of striking at greater distances, this is also true of the forces opposing them. Since the late 1950s the Soviet homeland has been threatened by long-range, carrier-borne bombers armed with nuclear weapons, and by submarines capable of launching their ballistic missiles from patrol areas in NATO-dominated waters. The nature of the maritime threat to the Soviet Union has therefore changed fundamentally since the 1920s, when the essential tenets of Soviet naval strategy and tactics were formulated.

From the early 1960s Soviet naval writers began to talk about 'blue zones' outside the traditional defensive boundaries of Soviet-dominated waters. Submarines powered by nuclear reactors, large missile-armed surface vessels, and land-based maritime bombers with greater endurance than their predecessors began to appear in numbers. Major Soviet naval exercises, in particular those of the OKEAN series, practised the engagement of hostile surface ships and submarines in open ocean areas such as the North Atlantic using Combined Arms tactics.

Gorshkov's 'Balanced Fleet'

Inevitably, the requirement to engage hostile naval forces at ever-greater distances from the Soviet land mass led some officers to question traditional Soviet Navy doctrine. It may be that operational shortcomings experienced during the major fleet exercises cast doubt on the effectiveness of Combined Arms operations under shore-based command in open ocean areas. Certainly there appears to have been some 'devolution' of responsibilities within the command chain during the 1970s, when major surface vessels began to appear equipped with extensive communications facilities which would enable them to exercise local command over groups of ships and submarines within their operational area.

The need for surface forces capable of sustained operations at sea brought with it a requirement for 'general-purpose' vessels closer in concept to those in service with Western navies. Earlier Soviet surface vessels, and indeed the submarines built from the late 1950s onwards, were 'mission-specific' (i.e., each was built for a relatively narrow mission, and was given only the features needed to perform that mission). This had created major problems when the nature of the Western threat changed, as it had done so dramatically when Polaris took over the US Navy's nuclear strike mission.

Many Soviet ship-types were designed as 'interceptors', and were therefore short on endurance. The Soviet Navy was also short of the underway replenishment vessels needed for sustained operations. Soviet surface ships operating in blue-water areas were increasingly dependent on surface-to-air missile systems to protect them against hostile strike aircraft, given the distances from friendly air bases at which they were being compelled to operate.

A fierce debate took place within the Soviet military hierarchy in the early 1970s, centred around a series of articles published under Gorshkov's name in the professional journal *Morskoy Sbornik* from 1972 to 1974, and entitled 'Navies in Peace and War'. These articles criticized the imbalance of ship-types in the current Soviet fleet, and postulated the need of 'general-purpose' forces able 'to carry out assigned missions in different circumstances'. Gorshkov appears to have won the argument on this occasion, and many of the ideas put forward in this series of articles were incorporated in his influential book *The Sea Power of the State*.

The culmination of this shift in Soviet tactical thinking was the laying down of the Soviet Navy's first through-deck carrier in 1982. The carrier would for the first time provide the Soviet Navy with its own integral tactical air support, enabling its surface ships to operate even in blue-water areas under a friendly 'unbrella'.

The fall of Gorshkov

The concept of the 'balanced fleet' was alien to traditional Soviet thinking. It implied: that naval tactics constituted a discipline independent of Marxist teaching; that naval construction and weapons programmes could be pursued without regard for their effect on those of the other armed services; and that the Soviet Navy could establish its own missions outside the land-based theatre command structure, and perform those missions by deploying self-contained Western-style task groups under Navy control.

All of this was heresy, and only a powerful figure with as many political friends as Gorshkov could have obtained some degree of acceptance for his proposals. They were, however, to be largely responsible for his downfall in 1986. The debate in Soviet professional journals which took place in the early 1980s, and to which we have already referred in respect of Soviet policy-making, also focused on the shortcomings of Gorshkov's tactical thinking.

On his accession to the post of Commander-in-Chief of the Soviet Navy, Admiral Chernavin restated firmly that the Combined Arms doctrine was central to Soviet naval operations, and stressed the importance of training in assuring the required degree of co-ordination. The major open-ocean exercises which had grown ever grander in scale under Gorshkov were immediately reined in; recent exercises have taken place almost exclusively within Soviet waters, with the emphasis on multi-dimensional Combined Arms attacks on surface ships simulating US Navy battle groups.

It is not without interest that Gorshkov, in the early part of his career, was a 'big-ship' sailor, whereas Chernavin is a 'submariner'. We appear to be on the threshold of major policy changes in the Soviet Navy – perhaps every bit as great as those which accompanied Gorshkov's own accession to the post of Commander-in-Chief in 1956.

A framework for analysis

In this extended introduction I have attempted to establish a broad framework for the analysis of Soviet submarine development in the post-war era. Many of the themes outlined above will recur in the following chapters, because they are fundamental to any consideration of the designed mission and technical capabilities of the various types of submarines constructed during the period.

The submarine occupies as central a place in Soviet naval thinking as does the tank in Soviet land strategy. Its importance is shown not only by the number of units that have been in service at any given time, but also by the degree of priority accorded to research and development in the area of underwater operations. Not only have large numbers of hulls been completed, but there has been a surprising number of different types, given that Soviet military hardware is generally characterized by uniformity and mass production. There are currently no fewer than eight classes of submarine either at the prototype stage or in full production, as compared with a maximum of two or three in any of the major Western navies. Some of these submarines bear little resemblance to their Western counterparts because they have been built in response to specific military requirements within a very different politico-economic system.

It is important not to allow Western perceptions to distort our view of Soviet submarine development. If we begin our analysis from the standpoint of viewing every Soviet submarine as a potential threat to Western security, we risk seeing each new development as a response to a perceived Western weakness, without taking into account Soviet perceptions regarding their own vulnerabilities. It is important in this respect to remind ourselves that every state has its own legitimate security requirements, and that these often have to be fulfilled by adopting strategies that are fundamentally different from those of others.

1. The Post-war Era

A pre-history

The priority accorded to submarine construction during the period preceding the Second World War was such that the Soviet Navy entered the war with a submarine force numerically as large as any in the world. This was a considerable achievement in view of the limited capacity of the post-Revolution industrial base and the intense competition for resources between the various branches of the Soviet armed services.

The 1933–7 Five Year Plan provided for the completion of no fewer than 355 submarines by the beginning of 1938, and although in the event it proved impossible to meet this target, the subsequent Five Year Plan was almost as ambitious, calling for the completion of 341 boats by early 1943.

The 1938–42 Five Year Plan was interrupted by Hitler's invasion of the Soviet Union in June 1941, and submarine construction was seriously affected by the rapid advance of the *Wehrmacht* on Leningrad, where the principal Soviet shipbuilding yards were located. Nevertheless, the Soviet Navy entered the Second World War with the impressive total of 202 submarines in service, of which six were ocean-going 'cruisers', 28 were 'large' boats of 950–1,250 tons displacement, 94 were 'medium' boats, and 78 were small coastal submarines.

The coastal boats of the 'M' (or *Malutka*) class were designed to be dismantled and transported from one operational area to another by rail, and were intended to protect Soviet coastal waters against incursions by hostile surface units. Their small size made them particularly well-suited to operations in the shallow and restricted waters of the Baltic and the Black Sea, but they were also deployed in large numbers in the Far East.

The medium submarines of the 'Shch' and 'S' classes constituted the backbone of the Soviet submarine force. Their primary mission was to defend the outer zone of Soviet sea-space in the four operational areas. It was envisaged that the medium boats would be allocated patrol sectors positioned across the estimated line of advance of enemy surface forces.

The third type of Soviet submarine was designed for long-distance operations in enemy waters, outside the Soviet sea-defence zones. Influenced by contemporary Italian submarines, the six 'D'-class boats were the first submarines designed and built in the Soviet Union. The design was not a conspicuous success, and they were subsequently subject to frequent repairs and modifications. They were followed by the 'L'-class minelayers, based on the design of the British *L55*, which had been sunk by Soviet destroyers off Kronshtadt in 1919, and was salvaged and refloated in 1928. Twenty-five units had been completed by July 1941, thirteen of which were built at Vladivostok for service in the Far East.

Apart from the three Soviet-designed boats of the 'P' class, which were a total failure and saw little active war service, the only other submarine in the 'large' category was the 'K' class, six of which had been delivered by the outbreak of war in July 1941 with a further six on the stocks at Leningrad. These boats were designated 'cruisers' by the Soviets (the 'K' is reported to stand for *Kreiserny*), and carried a powerful armament of guns and torpedoes (*see* table).

In the original design, provision was made for a small float-plane which was to be carried, partially dismantled, in a special hangar. Two prototypes of the SPL float-plane were duly completed and flown in 1933–5, but no aircraft-carrying submarine was ever built. The intention to incorporate such a capability in submarines of this class, however, suggests a mission not unlike that of the large 'cruiser' submarines of the Imperial Japanese Navy, which were designed primarily for reconnaissance against enemy surface forces. The heavy gun armament, comprising two single 100mm deck guns, would be particularly useful for operations aimed at disrupting enemy sea lines of communication, and the 'K's also had a formidable minelaying capacity which could be employed offensively in the approaches to enemy ports and harbours.

The numerical bias in favour of medium and coastal boats in the Soviet pre-war naval construction programmes illustrates the essentially defensive nature of Soviet submarine operations of the period. The Soviet Northern Fleet was relatively young – it had achieved 'fleet' status only in 1937, and continued to rely on the traditional Baltic shipyards around Leningrad for its support. This was made possible by the completion of the Stalin Canal which linked the Baltic with the White

Right: *The six units of 'D' class were completed in 1928–9. They were the first submarines designed and built for the Soviet Navy, and suffered numerous technical shortcomings. This is* Revolucyoner *(D4) shortly after completion. (Przemyslaw Budzbon collection)*

'LARGE' SUBMARINES

	'D' Class (Series I)	'L' Class (Series II)	'P' Class (Series IV)	'K' Class (Series XIV)
Displacement:	930t/1,350t	1,050t/1,325t	1,200t/1,870t	1,500t/2,100t
Length:	76m	81m	90m	98m
Beam:	6.5m	6.6m	8m	7.4m
Propulsion:	diesels, 2,600bhp; motors, 1,600hp; 14kt/9kt	diesels, 2,200bhp; motors, 1,050hp; 14kt/9kt	diesels, 5,400bhp; motors, 1,000hp; 18kt/8kt	diesels, 8,400bhp; motors, 2,400hp; 21kt/10kt
Armament:	eight 533mm TT 14 torpedoes one 100mm one 45mm	six 533mm TT 12 torpedoes 20 mines one 100mm one 45mm	six 533mm TT 10 torpedoes two 100mm one 45mm	ten 533mm TT 24 torpedoes 20 mines two 100mm two 45mm
Complement:	53	54	54	60

'MEDIUM' AND COASTAL SUBMARINES

	'Shch' Class (Series III)	'S' Class (Series IX)	'M' Class (Series VI)	'M' Class (Series XV)
Displacement:	580t/705t	840t/1,070t	160t/200t	280t/350t
Length:	57m	78m	38m	50m
Beam:	6.2m	6.4m	3.1m	4.4m
Propulsion:	diesels, 1,370bhp; motors, 800hp; 12kt/8kt	diesels, 4,000bhp; motors, 1,100hp; 19kt/9kt	diesel, 685bhp; motor, ?hp; 13kt/6kt	diesels, 1,600bhp; motors, 875hp; 15kt/8kt
Armament:	six 533mm TT 10 torpedoes one 45mm	six 533mm TT 12 torpedoes one 100mm one 45mm	two 533mm TT one 45mm	four 533mm TT one 45mm
Complement:	35	46	16	24

Sea, and through which the first Soviet submarines passed in 1933. Although the Northern Fleet was the only western fleet with free access to open waters, only two of the ocean-going 'K' class boats were deployed there at the outbreak of war, but the remaining four were quickly transferred from Leningrad via the Stalin Canal in August 1941, and spent the remainder of the war in northern waters. The bulk of the medium and coastal submarines were deployed with the Baltic and the Pacific Fleets, where they confronted the German and Japanese surface fleets respectively. The Black Sea Fleet had a level of provision slightly below that of the Baltic Fleet, while the infant Northern Fleet had only a handful of medium and coastal boats.

Soviet submarine technology

The influence of British and Italian designs on early Soviet submarine types has already been noted. Of even greater significance, however, was the assistance given to the Soviets by Germany during the period of the Weimar Republic.

German naval missions of the early 1920s brought plans of First World War submarines to show to their Soviet counterparts. Some German technology was also incorporated into the British-inspired 'L' and the home-grown 'P' classes, including twelve German MAN diesel units purchased direct from 1931 to 1935.

The design of the standard Soviet medium submarine of the pre-war period, the 'S' (or 'Sredny') class, was drawn up at the office of Ingenieurskaantor voor

Scheepsbouw in The Hague (Netherlands) in 1933, at a time when IKS was a joint-stock company owned by Krupp's Germaniawerft (Kiel), Deutsche Schiff und Maschinenbau AG (Bremen), and the German Navy. The Soviet design apparently served as a prototype for the German Type IA double-hull U-boat, which began construction shortly afterwards.

Under the Weimar Republic the German Navy, heavily constrained by the clause in the Treaty of Versailles prohibiting U-boat construction, saw Soviet interest in German designs as an opportunity to stay abreast of the latest technological developments in this category of vessel. The construction of Soviet submarines based on German designs effectively constituted 'building by proxy', and subsequent Soviet evaluation of the designs in service would, it was hoped, eventually provide the necessary feedback to develop newer, more capable types of submarine for the German Navy.

Once Germany threw off the shackles of the Versailles restrictions and embarked on the construction of a new generation of U-boats, the German Navy no longer had a vested interest in supplying submarine designs and technology to the Soviets. The new Chancellor, Adolf Hitler, viewed Soviet Russia as a potential military adversary. Nevertheless, the signing of the Nazi-Soviet Non-Aggression Pact in 1939 was accompanied by an agreement on the part of Nazi Germany to provide some military assistance, which included plans of its latest battleships and the transfer to the Soviet Navy of the incomplete 8in-gun cruiser *Lützow*. The Soviets also requested plans of the latest German U-boats, although this demand met with fierce opposition from the German Navy. There is no evidence to suggest that the Soviets were successful in gaining access to U-boat plans, but Jurg Meister, in his book on the Soviet Navy of this period,[1] has noted the external similarity between the Series XVI of the Soviet 'S' class and the German Type VIIC, the first of which was on the stocks when the agreement was signed.

Russia had been technologically backward under the Tsars, and had to rely heavily on foreign expertise for the construction of its fleets during the early part of the century. The October Revolution and the subsequent Civil War had all but destroyed what little infrastructure there was, and rebuilding the industrial base proved to be a gigantic task which had to be undertaken from a position of almost total political isolation. Although the submarines in service with the Soviet Navy at the outbreak of the Second World War represented the best Soviet technology of their day, it was only natural that technical problems should have been experienced in operations. The small size of the 'M' class boats led to buoyancy problems, and the firing of

Left: *The 'medium' boats of the 'Shch' and 'S' classes constituted the bulk of the Soviet submarine fleet during the Second World War. This is S56 of the Series XI-bis, preserved ashore at Vladivostok since 1972. (Przemyslaw Budzbon collection)*

Left: *The tiny 'Malutkas' were designed for coastal defence, and could be transported by rail between the various fleet areas. This is M30 of the Series XII-bis, photographed at a post-war naval review. (Przemyslaw Budzbon collection)*

Left: *The 'cruiser' submarines of the 'K' class were the largest, and arguably the most successful, of the Soviet wartime submarines. The submarine depicted is one of the later units of the class. Note the single 100mm guns trained fore and aft. (Przemyslaw Budzbon collection)*

a torpedo frequently caused them to surface involuntarily. The 'D' class were dogged by unreliable flooding valves which often stuck in the open position, and their sealed battery compartments made them difficult to maintain.

The Soviets attempted to resolve these defects by progressive modifications to those boats already in service, and by frequent redesign of the types under construction. The 'L' class comprised four different series; the 'Shch' class six; the 'M' class four, and the 'S' class three. Each successive sub-group featured increases in size, and improvements in armament and performance.

The Soviets also experienced technical problems with their torpedoes, which proved unreliable in service. These defects, when combined with deficiencies in firing doctrine, which stressed the importance of releasing weapons from close to their maximum range (in order to safeguard the submarine), resulted in a preference on the part of many Soviet submarine commanders for using deck guns.

Wartime operations

The Soviet submarine strategy projected during the 1930s as the basis for submarine design and construction policy had to be heavily modified in the light of subsequent events. When the *Wehrmacht* swept along the Baltic coast and placed Leningrad under siege, the available sea-space contracted to such an extent that the Soviet Navy was effectively penned inside its bases within a tight blockade imposed by a combination of anti-submarine surface forces and minefields. Soviet submarine operations were therefore directed primarily against breaking that blockade in order to disrupt enemy resupply operations. Soviet submarine commanders showed considerable courage and initiative in breaking through the formidable German anti-submarine barriers in the Gulf of Finland, but had few successes against Axis merchant vessels, and experienced heavy losses in the process.

The German occupation of Norway, however, created a more favourable operating regime for the submarines based in the Arctic, which by August 1941 included all six of the ocean-going 'K' class submarine cruisers. These boats were more successful, conducting mining operations in the Norwegian fiords and landing special reconnaissance teams. There were also a few successes against surface shipping, although again these were limited due to technical problems and deficiencies in operational doctrine.

The more open situation prevailing in northern waters allowed experimentation with submarine tactics to an extent that was not possible in the other theatres. Various deployment schemes were apparently tried out in order to evaluate the relative effectiveness of static ('positional') and mobile ('cruising') formations. The Soviet Navy is also known to have employed a tactic known as the 'hanging curtain', in which a group of submarines was deployed along the coast around the North Cape in a series of narrow, adjacent patrol areas, with supporting aircraft and surface ships. The submarines patrolled on the surface until reconnaissance information regarding the movement of German coastal convoys was transmitted to them from the controlling post ashore. The 'curtain' was then lowered and the submarines would lie in wait for their targets.[2] This constitutes a classic example of Soviet Combined Arms tactics, and is not so far removed from the sort of defensive deployment pattern one might expect of modern Soviet diesel boats.

In the Black Sea, submarines appear to have been used extensively for covert operations and for the support of amphibious landings. Submarines were used to insert reconnaissance teams, to carry supplies to isolated shore positions, and even for gunfire support. Similar operations were extended to the Baltic coast in the latter part of the war, when the German Army was in full retreat.

Victory and beyond

The Soviet Navy ended the war with a fleet of 173 submarines, of which some 40 per cent were small coastal boats. There was also now a substantial body of personnel with operational experience in the three western fleets. The latest types of Soviet submarine, the 'K'-class ocean-going cruisers and the 'medium' boats of the 'S' class, had proved relatively successful, and preliminary designs had been drawn up for their successors.

Clearly, access to German technology would have a significant impact on subsequent submarine developments. The final months of the war in Europe, however, were to be marked by an unprecedented scramble for German expertise among the major Allied powers which was to result in a less-than-even distribution of the products of German military science. Well before the final surrender the relationships between the three major Allies had been clouded by suspicion regarding post-war foreign policy ambitions. Their armed forces therefore entered German territory with 'shopping lists' provided by their respective Intelligence services.

The resulting acquisitions largely reflected geography, with the German research establishments of the east falling into Soviet hands, and those of the west falling to the British or Americans. Prior agreements to distribute the spoils evenly were not always honoured.

Left: *Early military disasters in European Russia compelled the Soviet Navy to shift the focus of its submarine operations to the Arctic. Here a 'medium' submarine of the 'Shch' class comes alongside a 'K'-class cruiser. (TASS)*

In general terms, the Soviet Union benefited considerably from its acquisition of Peenemünde, where the V-1 and V-2 rockets had been developed and tested, but lost out with regard to advanced submarine propulsion systems. The Soviets were therefore well to the fore with the deployment of cruise and ballistic missile systems aboard submarines during the 1950s, but in spite of a keen interest in closed-cycle propulsion their attempts to develop an operational system met with total failure.

German war spoils

In accordance with the Potsdam agreement, the Soviet navy received a number of former Axis submarines as its share of war spoils. There were two ex-Italian 'medium' boats, the former *Marea* and *Nichelio*, but of greater interest to the Soviets were the ex-German submarines, which represented the full range of types completed for operational service during the Second World War. From 1946 to 1947 four Type VIIC 'medium' U-boats, a single Type IXC, four of the advanced Type XXIs, and a single Type XXIII coastal U-boat were commissioned into the Soviet Navy. They were subsequently used for trials and evaluation.

The Soviets also captured fourteen incomplete Type XXIs in an advanced state of construction, together with a large number of prefabricated sections, when they entered Danzig in 1945. They gained access to the Walter design office at Blankenberg, where a full-scale wooden mock-up of the projected Type XXVI U-boat was located, together with the Walter turbine plant and two complete turbines. However Dr Walter himself was sought out and captured, together with his design staff, by the British at Kiel. The British rapidly shipped out all the relevant documents and machinery parts and transferred them to the Vickers shipyard at Barrow-in-Furness. They also salvaged a number of Walter prototype U-boats, including *U 1406* and *U 1407*, which had been completed only just before the war ended. The former vessel was handed over to the Americans, but *U 1407* was towed to Barrow and completely refurbished under the supervision of Dr Walter and his staff. Trials were conducted from 1947 until 1949, and experience with this boat was incorporated into the design of the Royal Navy's experimental submarines *Explorer* and *Excalibur*.

In spite of having access to Dr Walter and his design team, plus a complete Walter U-boat, it was to take the British ten years to complete their own prototypes. This gives an indication both of the difficulty involved in assimilating such advanced technology, and of the scale of the technical problems experienced in developing an operational Walter turbine. It is therefore hardly surprising that the Soviets, who were in possession only of plans, a wooden mock-up and a pair of experimental turbines, should have had so little success in their own attempts to develop closed-cycle propulsion systems.

The post-war programme

In 1948 the US Office of Naval Intelligence estimated the current strength of the Soviet submarine fleet at 281 units, of which thirteen were elderly boats and no fewer than 103 were small coastal types. But in that same year a Soviet admiral spoke of a 20-year programme to produce 1,200 submarines of new design. This not unnaturally provoked considerable alarm in the West, which saw in these figures a potential threat to the North Atlantic sea lanes even greater than that posed by the German U-boats at the peak of their operations in 1942.

Western fears were exacerbated by the knowledge that the Soviet Navy now had access to the best German submarine technology. Analyses of future Soviet capabilities were therefore based on a 'worst-case' scenario. The ONI estimated that as early as 1951 the Soviet Navy would have in service 356 submarines, including 26 Type XXI equivalents and 60 snorkel-equipped boats with high surface speed. It was thought that the Soviets would complete all the unfinished Type XXIs they had captured at Danzig, and would then embark on series production of 'carbon copies' in their own shipyards. It was estimated that the capacity of Soviet shipyards was currently 20–30 submarines per year, but that this would rise within five years to the German peak rate of 30 per month!

The 1950–51 edition of *Jane's Fighting Ships* reported the construction of 120 submarines of new design in Soviet shipyards. Four distinct types were stated to be in production, all of which were to be equipped with snorkels: an improved 'K'-class ocean-going type, a new fast minelayer, the German Type XXI, and the German Type XXIII. This statement subsequently proved to be without foundation, but it provides a good illustration of Western thinking of the time, dominated as it was by fears of a German-style U-boat campaign conducted by the Soviet Navy against the North Atlantic shipping lanes.

Unfortunately, these initial assessments by the Western Intelligence services were to colour analysis of Soviet submarine developments for at least the next twenty years, and were to result in a series of errors regarding the capabilities, projected missions, and building rates of the various submarine types which subsequently entered service.

Reconstruction

Soviet perceptions were somewhat different. It was no longer the *Wehrmacht* that posed the major threat to Soviet security, but the victorious navies of the Western Allies. The US Navy in particular, boosted by the largest wartime naval construction programme in history, was capable of unprecedented power projection operations employing large amphibious fleets supported by carrier task forces.

Moreover, the maritime thread had changed not only in magnitude but in direction. The Baltic and the Black Sea, which traditionally had been the focus of Russian naval operations, and in which all the major shipbuilding yards and naval bases were located, were both enclosed seas which, in the aftermath of the Second World War, were dominated by the Soviet military presence. The areas of Soviet vulnerability were the two 'open flanks' in the north and in the Far East.

This new strategic situation prompted a radical reappraisal of the overall structure of the Soviet Navy, of the types of vessel required, and of the capacity and location of shipbuilding and maintenance facilities.

Stalin, with the support of the more conservative elements of the Soviet Navy, appears to have favoured the construction of an ocean-going fleet capable of challenging the navies of the Western Allies on the high seas, and in about 1948 is reported to have sanctioned a large programme of surface warships, which included aircraft carriers and battlecruisers. However, the Soviet shipbuilding industry, which had been devastated by the war, was incapable of delivering such an ambitious programme. The first post-war Five Year Plan was therefore largely devoted to the reconstruction of the traditional shipyards, and the further development and expansion of the relatively new shipyards at Severodvinsk in the Arctic and Komsomolsk in the Far East.

The policy of 'infrastructure first, production later', was a well-established Socialist doctrine. Shipyard No. 402 at Severodvinsk (known until the late 1950s as the Molotovsk Shipyard) was begun during the mid-1930s following the completion of the Stalin Canal connecting the Baltic and the White Sea, and was intended to reduce the dependence of the new Arctic Flotilla (later 'Fleet') on the traditional shipyards around Leningrad. The massive undercover construction hall, completed with forced labour, measured 335m by 137m and was capable of accommodating the hulls of two large battleships side by side. Initially the shipyard remained dependent on components and material transported from Leningrad and Nikolayev, and this prevented it from reaching its full capacity during the 1940s. By the 1950s, however, Shipyard No. 402 was building large conventional cruisers of the *Sverdlov* class and *Skory*-class destroyers. It would become increasingly important for the construction of large submarines for the Soviet Navy.

Shipyard No. 199 at Komsomolsk (also known as Leninskaya Komsomola) was established during the 1930s as a parallel development, designed to provide for the planned expansion of the Soviet Pacific Fleet. Although limited in the maximum size of hull it could construct by its position 280 miles above the mouth of the River Amur, the Komsomolsk Shipyard laid down cruisers and destroyers during the immediate post-war period, and was soon to become a major submarine building yard.

The Sudomekh Shipyard in Leningrad, formerly the Neva Admiralty Dockyard, had been developed as a specialist submarine building yard during the 1930s. Although extensively damaged during the Second World War, it was quickly rebuilt during the late 1940s, when equipment removed from German submarine construction yards was installed. A number of 'S' class hulls laid down during the hostilities were completed, and construction then switched to one of the major new classes of submarine, the 'Zulu'. The Sudomekh Yard was to become closely associated with developmental types, and in particular with advanced propulsion systems.

Perhaps the most remarkable of the new submarine building yards was the inland Krasnaya Sormovo Shipyard at Gorky. Situated on the River Volga at its confluence with the River Oka some 320 kilometres east of Moscow, the yard was developed during the Second

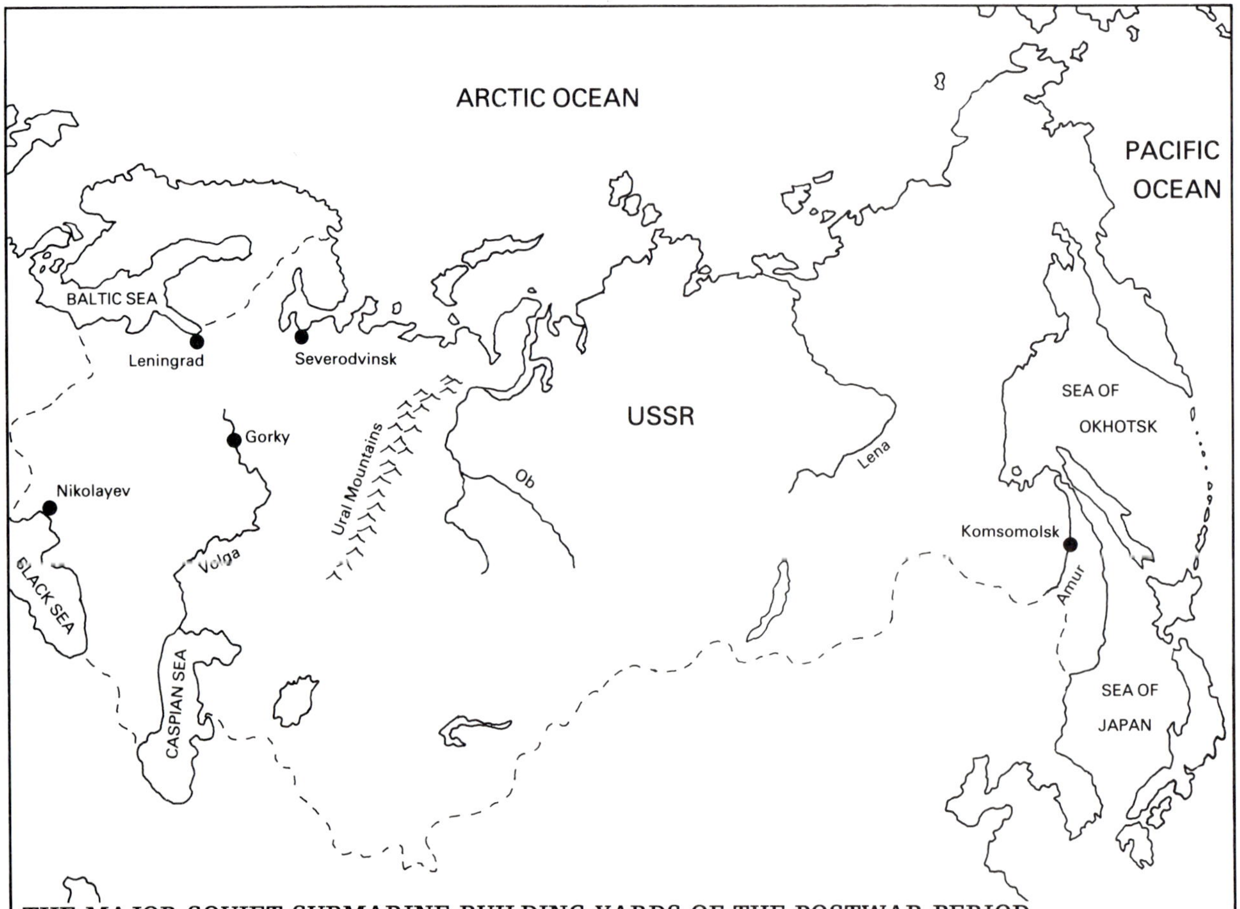

THE MAJOR SOVIET SUBMARINE BUILDING YARDS OF THE POSTWAR PERIOD

Severodvinsk:
Molotovsk Shipyard

Gorky:
Krasnaya Sormovo Shipyard

Leningrad:
Ordzhonikidze Shipyard
Marti Shipyard
Sudomekh Shipyard

Komsomolsk:
Leninskaya Komsomola Shipyard

Nikolayev:
Marti Shipyard

World War as a safe haven for construction beyond the reach of German bombers. Initially the yard prefabricated sections of 'M'-class coastal submarines which were then shipped out on barges to the coastal shipyards for assembly. During the latter part of the war, however, the Gorky Shipyard built complete 'S'-class hulls. Because its construction facilities emerged from the war undamaged, the Krasnaya Sormovo Shipyard was well-placed to begin construction of the 'S'-class replacement during the late 1940s. Although initially it specialized in mass-production techniques, the Gorky Shipyard was eventually to build large and complex nuclear-powered boats.

The traditional yards

Three older, well-established shipyards were to continue to play a part in post-war submarine construction. The Ordzhonikidze Shipyard, Leningrad, had been the major Soviet submarine building yard before the Second World War, but suffered extensive damage during that conflict and had to be completely rebuilt. Naval construction during the immediate post-war period was limited to six cruisers of the *Sverdlov* class, and the building of submarines was resumed only for a short period during the mid-1950s. Following reconstruction, the yard apparently reverted to its pre-Revolution name, and was subsequently known as the Baltic Shipyard.

The Martin Shipyard, also at Leningrad, was the other major pre-war submarine building yard in the Baltic. It too suffered extensive damage during the war, but was rebuilt to enable it to play its part in the post-war surface ship construction programme, Like the Ordzhonikidze Yard, it subsequently reverted to its pre-Revolution name and became the Admiralty Shipyard. It would play an important part in submarine construction, but only after extensive rebuilding in the early 1960s.

The third major pre-war submarine construction yard was the Marti Yard, Nikolayev, on the Black Sea, which completed large numbers of 'L'-class and 'Shch'-class boats, as well as building and assembling the tiny 'M'-class coastal submarines. Although extensively damaged during the war, the yard was quickly rebuilt, and played a major part in the 'Whiskey' construction programme, launching its first hull as early as 1951.

The importance of these traditional Baltic and Black Sea submarine building yards had inevitably declined as a result of wartime damage and the establishment of large-scale submarine production at the inland Gorky Shipyard. There was, therefore, a deliberate reorientation of their role towards the construction of the large surface ships which figured so prominently in Stalin's first post-war programme. With the termination of the 'Whiskey' series in the mid-1950s, submarine construction would be focused exclusively on the 'new' specialist submarine building yards of Sudomekh, Gorky, Komsomolsk and Severodvinsk, which offered a much-improved geographical distribution as well as the latest construction facilities.

The new submarines

Once the reconstruction of the Soviet shipyards had been completed, new types of submarine were laid down. It was initially assumed by Western defence analysts that the new Soviet boats would be carbon copies of German designs of the late-war period. The 'Whiskey' was for many years described by reference sources as a Soviet version of the Type XXI. This was a serious error which served to distort Western perceptions of the capabilities and designed mission of these submarines.

The German Type XXI, considered revolutionary by the Western Allies, was a large submarine of 1,620 tons surfaced displacement, designed for commerce raiding. Built around a 'figure-of-eight' pressure hull, it incorporated an additional bank of batteries amidships which provided the necessary power for prolonged high-speed underwater operations. A snorkel allowed the submarine to recharge its batteries while running just below the surface.

The pressure hull of the Type XXI was enclosed within a cigar-shaped outer casing designed to reduce resistance at high speed, and there was a long streamlined fin. The fine lines aft precluded stern tubes, so all six of the torpedo tubes were in the bow. Abaft the tubes was a capacious handling room which could hold an additional seventeen anti-ship torpedoes.

The Soviet 'Whiskey' bore little resemblance either externally or internally to the Type XXI. It was a much smaller boat, with a surfaced displacement of only 1,050 tons, and was based on a conventional cylindrical pressure hull. In keeping with previous Soviet practice it employed direct drive, the diesels being paired with the electric motors on each of the two shafts. Motor power was therefore limited by diesel power, so there was little to be gained by fitting additional banks of batteries. (In contrast, US Navy warbuilt fleet boats, which had diesel-electric drive, were easily converted to Type XXI performance simply by uprating their battery capacity and installing more powerful electric motors.)

Although of double-hull design, the 'Whiskey' had prominent 'saddle tank' bulges in her outer casing. The fin was also far less streamlined that that of the Type XXI; in early boats of the class there was a prominent

forward extension for a small twin AA gun turret, and abaft the fin there was a single 100mm deck gun, suggesting a continued predilection on the part of Soviet submariners for fighting on the surface. Finally, the six torpedo tubes were distributed fore and aft in the conventional way, with torpedo rooms serving both sets of tubes. Four of the tubes were in the bow and two in the stern, and in contrast to the heavily armed German boats, which were designed for prolonged anti-shipping patrols, there were only six reloads (one per tube).

More revealing is a comparison between the 'Whiskey' and the warbuilt Soviet 'S'-class medium boats. Length and beam were virtually identical, and the 'S' hull-form was based on the same double-hull saddle tank design. The twin diesels of the 'Whiskey' direct drive propulsion system were almost certainly derived from the model installed in the 'S' class, and brake horsepower was identical. The same layout of torpedo tubes and handling rooms was adopted, and a total of twelve torpedoes was carried by both types. Both the 'S' and the early 'Whiskeys' mounted a single 100mm deck gun, together with smaller guns for defence against aircraft.

The only significant difference in performance lay in the higher submerged speed of the 'Whiskey' (13.5 knots as compared with 9 knots in the 'S'-class), which was achieved by more than doubling motor power and by the installation of more powerful batteries. The latter, together with increased oil fuel bunkerage to improve endurance, was presumably largely responsible for the greater draught of the 'Whiskey' (4.8m as opposed to 4.1m) and a 200-ton increase in displacement.

Greater endurance was required because of the change in the nature of the perceived threat. The 'Whiskey' had to be suitable not only for relatively short-range operations in the Baltic and the Black Sea, but also for the defence of the more open waters in the north. Patrols would take in not only Arctic waters, but also the northern reaches of the Norwegian Sea. The increase in draught and displacement over the warbuilt

Above: *A 'Whiskey V' running on the surface in the English Channel during the late 1960s. Note the characteristic colour scheme of the period, comprising a white pennant number, together with a black line topped with white at the base of the fin and a similar colour scheme at the surfaced waterline. (Skyfotos, 1968)*

'WHISKEY V' CLASS

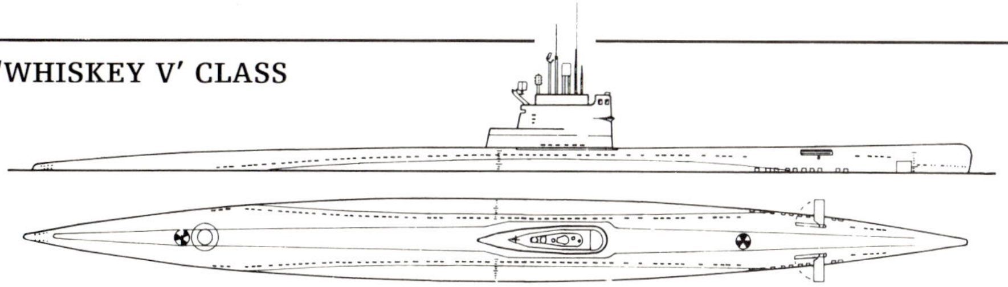

'WHISKEY' CLASS

Soviet designation: Podvodnaya Lodka (PL) (Submarine)

Construction:
Built at Baltic Shipyard (Leningrad); Krasnaya Sormovo Shipyard (Gorky); Komsomolsk-na-Amur; Marti Shipyard (Nikolayev).
236 units completed 1950–7: one in 1950, four in 1951, sixteen in 1952, 23 in 1953, 48 in 1954, 62 in 1955, 63 in 1956, 19 in 1957.

Displacement:
Surfaced: 1,050 tons
Submerged: 1,350 tons

Dimensions:
Length: 76m (249ft)
Beam: 6.3m (21ft)
Draught: 4.8m (16ft)

Propulsion:
Diesel/electric drive on two shafts; two Type 37D diesels, 4,000bhp for 17 knots surfaced; two electric motors, 2,500hp for 13.5 knots submerged.

Complement 50

Weapons:
Some units completed with 100mm deck gun and twin 25mm AA mounting, some with twin 25mm AA only (removed by late 1950s); four bow tubes, two stern tubes, all of 533mm (21in) diameter, for anti-ship torpedoes; maximum capacity twelve torpedoes.

Sensors:
HF passive sonar; search and attack periscopes; Snoop Plate surface surveillance radar; Stop Light ESM; Quad Loop DF.

Communications:
HF and VHF radio.

'S' class provided both the necessary bunkerage and improved seaworthiness – an important consideration for submarines which would continue to spend a large proportion of their patrol time on the surface.

There can be little doubt that the Soviets took full opportunity of the 'breathing space' afforded by shipyard reconstruction during the immediate post-war period to modify the designs drawn up during 1944 and 1945 in the light of their own wartime experience, and to incorporate into the new submarines as much German technology as possible. It is equally clear, however, that both in design and concept the 'Whiskey' is essentially derived from earlier Soviet medium submarines. Some of the sub-systems developed during the immediate post-war period undoubtedly benefited from German 'know-how', but the primary mission of these submarines, that of protecting Soviet sea-space from hostile intruders, remained unchanged.

Moreover, there are indications that Western analysts have overestimated the speed with which the Soviets would have been able to assimilate the advanced technology they acquired from the Germans. The devastation wrought during the war years on Soviet industry in general, and on the shipbuilding industry in particular, was immense. These industries could not be reconstructed overnight, and it was to be many years before the Soviets were able to assimilate and to develop for their own purposes concepts which in many cases the Germans themselves had found difficulty in bringing to fruition.

One particularly surprising feature of the submarines built by the Soviets during the early 1950s was

Above: *The drawing shows the final 'Whiskey V' configuration.*

Right: *A 'Whiskey V' in the Danish belt, Note the retractable forward hydroplanes, and the prominent 'saddle tanks'.*

INTERIOR PROFILE OF A 'WHISKEY V'

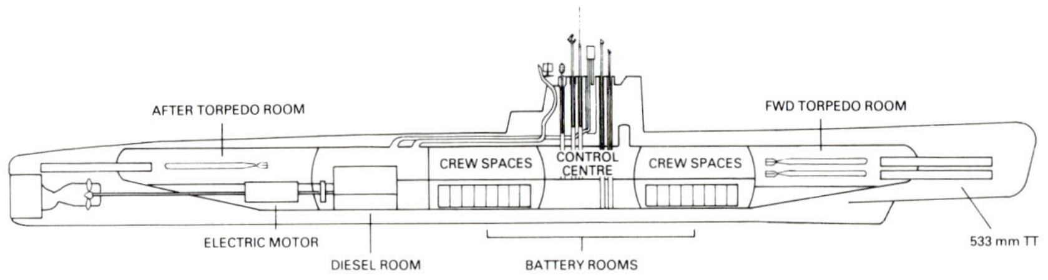

the absence of the snorkel, which was fitted only in later boats of the first generation. Different operating requirements, which did not envisage lengthy transits through enemy-dominated waters, may have been partly responsible for this omission. However, the variety of snorkel configurations adopted for the classes of submarine that entered service in the late 1950s suggests that the Soviets may have experienced problems in developing a workable system.

The 'Zulu'

The other major submarine type begun during the early post-war period was the 'Zulu', a large ocean-going boat of 1,900 tons surfaced displacement. The first unit was probably laid down in 1951 at the Sudomekh Shipyard – two years after construction of the 'Whiskey' had begun at Gorky.

The original design dated back to 1944–5, when the Soviets began work on a successor to the 'K' class, but the plans were subsequently put on ice pending a thorough examination of the German Type XXIs transferred to the Soviet Navy in 1947, and of the plans for even more advanced U-boats with closed-cycle machinery which fell into Soviet hands at the end of the war. The Sudomekh Shipyard, which had been hastily re-equipped during the post-war reconstruction programme with equipment removed from German submarine building yards, proved to be the natural choice as lead yard for the new Soviet boats.

The overall dimensions of the 'Zulu' were already fixed by the plans drawn up in 1944–5, and were much closer to those of the warbuilt 'K'-classs cruisers than to those of the German Type XXI, which was significantly shorter. The 'K's had proved relatively successful in service, and the Soviets were clearly reluctant to abandon completely an established design. They also wanted a large submarine capable of operating in the North Atlantic and off the East Coast of the United States. The nature of Soviet geography dictated that a submarine designed for such a mission would need to be larger even than the Type XXI, as the Soviet boats would have to deploy from the Arctic, not from the bases in the North Sea and Norway available to the German U-boats. The additional distance which would need to be covered before the Soviet submarines reached their North Atlantic patrol areas was in the order of 1,000 nautical miles.

Moreover, differences in Soviet tactical doctrine, which was founded on shore-based central control, meant that the 'Zulu' would spend more time on the surface than its German counterpart, and large size equated with the seaworthiness essential for lengthy patrols. Rather than conducting independent operations, it was envisaged that the 'Zulu' would provide early reconnaissance for Soviet forces operating within the sea-defence zones, enabling those forces to be disposed in such a way as to intercept an enemy fleet once it entered Soviet sea-space. Soviet submarine tactics therefore stressed the importance of continuous two-way communication, which was possible only if the submarine operated on or just below the surface.

Nevertheless, in technical if not in conceptual terms, the 'Zulu' was much more heavily influenced by the Type XXI than the 'Whiskey'. Externally, the long cigar-shaped hull-form closely resembled that of the German boats. This was also true of the stream-lined fin of the early 'Zulus' which, in the manner of the Type XXI, had twin AA gun turrets faired into its after end at the upper level (*see* drawings). These similarities became less marked when the fin was remodelled to incorporate a snorkel and a revised outfit of sensor masts during the late 1950s.

There were reported attempts to develop closed-cycle Walter turbines to power the outer two shafts of the 'Zulu' (the centre shaft would have had conventional diesel electric propulsion). A similar three-shaft propulsion system had been planned for the German Type XVIII U-boat from which the conventionally powered Type XXI had been developed. Two submarines of this type had been ordered from Germaniawerft, Kiel in January 1943, but they were in advance of development

Right: *Overhead of a 'Zulu', showing the distinctive cigar-shaped hull-form derived from the German Type XXI, and which was to be a feature of all Soviet diesel boats until the 1970s. (MoD, 1981)*

and were subsequently cancelled in favour of a 'crash' programme of Type XXIs.

It remains unclear whether Walter turbines were actually installed in the early boats of the 'Zulu' class. One source,[3] states that the first twelve units were so fitted, and that developmental problems with the turbines were responsible for the delays experienced in getting the submarines into service. Certainly construction of the 'Zulu' appears not to have proceeded as smoothly as the Soviets would have hoped. By the time the submarine was finally accepted by the Soviet Navy the Walter turbines reportedly planned for the outer shafts had been replaced by conventional diesel/motor units. However, the electric motors installed were even more powerful than those developed for the 'Whiskey', and maximum speed submerged was estimated at 16 knots, which placed underwater performance on a par with the Type XXI.

The 'Zulu' is generally credited with six bow torpedo tubes and four stern tubes, an arrangement similar to that of the 'K' class. The latter, however, had only two internal stern tubes, the remaining two being paired in a trainable mounting outside the pressure hull. The stern section of the 'Zulu' would therefore have to be

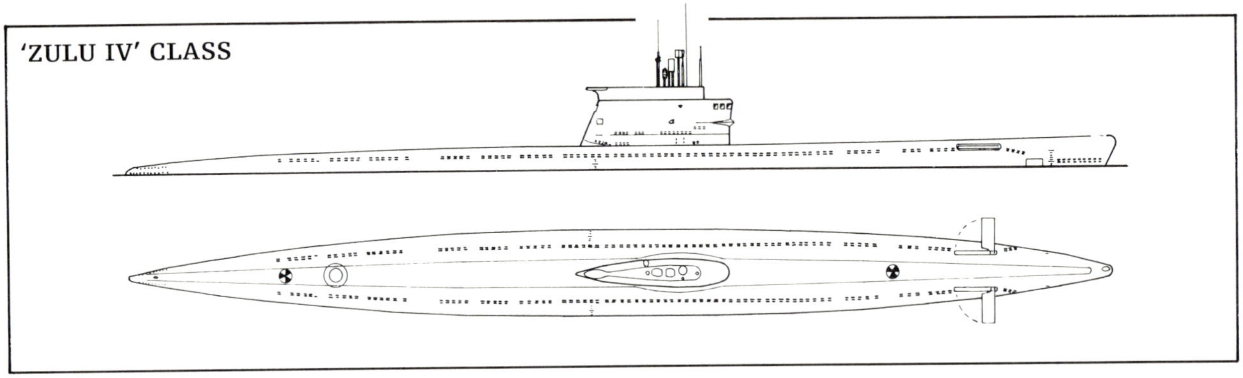

Right: *The 'Zulu' incorporated many of the features of the German Type XXI. Even though it was the only Soviet post-war submarine with sufficient range and endurance for commerce raiding, it appears more likely that its primary mission was reconnaissance, and that its targets would have been surface warships rather than merchant vessels. (US Navy, 1974)*

fuller than that of the Type XXI, which was fitted only with bow tubes, and the after tubes would require a second handling room.

Total torpedo capacity was about the same, 22–24 torpedoes being the preferred estimate (the 'K' class carried a full load of about 24). There was no separate provision for mines. In the 'K' class, twenty EM Type mines could be carried in the central ballast tank. By the time the final 'Zulu' designs were drawn up, however, the Soviets had developed influence mines based on German models, which could be launched from a standard 533mm torpedo tube. Two such mines could be carried in place of a single torpedo. The 'Zulu' could therefore carry a theoretical maximum of 44–48, although clearly a full load of mines would leave the submarine incapable of making torpedo attacks on any surface vessel it might encounter.

The medium submarines of the 'Whiskey' class also benefited from this development, which explains the absence of a specialist minelaying submarine in the post-war programme.

It is reported that some early 'Zulus' may have been fitted with a single 100mm deck gun forward of the fin, although in the absence of photographic evidence this is difficult to confirm. The deck gun was presumably justified by wartime experience with the 'K' class, but if fitted it would undoubtedly have created substantial drag on a hull otherwise designed for high underwater performance.

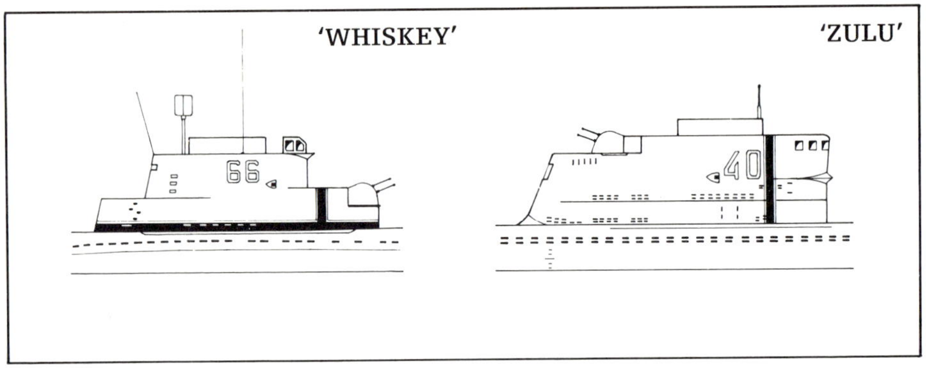

Construction

An estimated 236 'Whiskeys' were completed from 1950 to 1957. During the same period only 26 'Zulus' were built. This comparatively low figure can be explained in part by the difficulties initially encountered in the construction of what was in effect a much more technically advanced type of submarine. However, these difficulties do not provide the complete explanation, which can only be that the Soviet post-war programme did not envisage mass production of the

Upper left: *These two drawings of the conning towers of the 'Whiskey' and 'Zulu' are based on published photographs dating from the late 1950s. The 'Whiskey' had a projection forward of the conning tower on which a twin enclosed 25mm AA mounting was located. The 'Zulu' had a similar enclosed mounting at the after end of the conning tower, in the manner of the German Type XXI. Both types subsequently had their guns removed, and the upper section of the fin was built up to house a new array of sensor masts. Note the recognition markings, which were carried by all Soviet submarines during the 1950s.*

Left: *Endurance was a key factor in the 'Zulu' design. Here, submariners under the command of Vice-Admiral A. Sorokin parade following the Soviet Navy's first world submarine circumnavigation. (TASS, 1966)*

'Zulu' on the same scale as the 'Whiskey'. Given that the tactical relationship between the 'Zulu' and the 'Whiskey' was similar to that between the ocean-going 'K' class and the medium 'S'- and 'Shch'-class boats of the pre-war period, there would seem little justification for altering the numerical ratio. The only new factor was the increased importance of the Arctic flank, which favoured the larger submarine. There would, however, still be a requirement for large numbers of medium boats in the Baltic, the Black Sea and the Pacific.

Moreover, the 'Whiskey' was a larger and more capable submarine than the pre-war medium boats, and could be easily mass-produced at the Krasnaya Sormovo building facility at Gorky. Gorky is reported to have built 145 'Whiskey'-class hulls, more than 60 per cent of the total number, with no fewer than 40 being completed in the peak year of 1955. The other major contributor to the programme was the Marti Shipyard, Nikolayev, which built 65. The Ordzhonikidze Shipyard, Leningrad, and the Komsomolsk Shipyard in the Far East were relatively minor contributors, being credited with fourteen and eleven hulls respectively.[4]

The Sudomekh Shipyard, Leningrad, was the lead yard for the 'Zulu', the first of which was launched in 1952. Sudomekh was to complete a further 17 boats before production of the 'Zulu' was terminated in 1955. The only other Soviet shipyard involved in the programme was No. 402 Shipyard, Severodvinsk, which built eight. This was a logical step, as the 'Zulu' was designed primarily for operations in the Arctic, and establishing production at the Severodvinsk Shipyard would ensure that the Northern Fleet was able to maintain the submarines once they had been deployed to that area. It also helped to establish the expertise necessary for the construction of larger and more advanced submarines in the late 1950s.

The extent to which Western Intelligence sources misread the direction of the Soviet postwar submarine programme is suggested by persistent reports in *Jane's Fighting Ships*, beginning with the 1955 6 edition, that the 'Zulu' was being built at the rate of 18–20 boats per year. The 1958–9 edition credited the Soviet Navy with 60 'Zulus' and only 100 'Whiskeys'. Four more years were to elapse before these figures were corrected and the true balance between the two types was revealed to the public.

Some other problems

The difficulty of obtaining accurate intelligence on the Soviet postwar construction programme extended to the precise configuration of the submarines.

The sketchy nature of the photographic coverage of the 'Whiskey' and 'Zulu' classes during the early and mid-1950s leaves a number of questions unanswered.

In particular, some of the published drawings depicting fin configurations correspond only loosely with published photographs, while others depict configurations for which there appears to be no photographic evidence. Moreover, the 'type' designations ascribed to the numerous variants ('Whiskey I', 'Zulu III', etc.) appear to be somewhat arbitrary, and do not reflect the logical chronological sequence of the respective modifications. Inconsistencies in type designation between the various reference sources do not add to one's confidence in the accuracy of the information presented.

The true sequence of 'Whiskey' modifications appears to be as follows. Early boats had extensions fore and aft of the fin, with a twin 25mm gun turret incorporated into the forward extension. A single 100mm deck gun was fitted abaft the fin. The deck gun was not fitted in all boats; those initially fitted with it can be distinguished from later boats by the prominent flat 'gun deck' abaft the fin, which was retained even after the guns were removed.

Following the removal of the deck gun, some boats also had their 25mm AA gun turret removed. These modifications were finally superseded by a completely rebuilt fin, in which a snorkel was incorporated. The forward extension which had previously housed the 25mm gun turret was completely eliminated, and the Quad Loop direction-finding aerial, which had previously been mounted on a retractable mast, was relocated on the angled snorkel exhaust at the after end of the fin. By the 1960s all 'Whiskeys' were being refitted to this standard, which is generally known as 'Whiskey V'.

The early configurations of the 'Zulu' are less clear. Some sources credit the early boats with a single 100mm deck gun forward of the fin, others with a twin 57mm mounting. However, there appear to be no published photographs showing either, and there is no

'ZULU' CLASS

Soviet designation: Podvodnaya Lodka (PL) (Submarine)

Construction:
Built at Sudomekh Shipyard, Leningrad; Shipyard No. 402, Severodvinsk.
26 units completed 1952–5: one in 1952, eight in 1953, eight in 1954, nine in 1955.

Displacement:
Surfaced: 1,900 tons
Submerged: 2,350 tons

Dimensions:
Length: 90m (295ft)
Beam: 7.5m (25ft)
Draught: 6m (20ft)

Propulsion:
Diesel/electric drive on three shafts; three Type 37D diesels, 6,000bhp for 18 knots surfaced; three electric motors, 5,300hp for 16 knots submerged.

Complement 70

Weapons:
Some units completed with 100mm deck gun and twin 25mm AA mounting, or with two twin 25mm AA (removed by late 1950s); six bow tubes, four stern tubes, all of 533mm (21in) diameter, for anti-ship torpedoes; maximum capacity 22 torpedoes.

Sensors:
HF passive sonar; search and attack periscopes; Snoop Plate surface surveillance radar; Stop Light ESM; Quad Loop DF.

Communications:
HF and VHF radio.

'gun deck' similar to that abaft the fin of the early 'Whiskeys'. The first photo published in *Jane's* (1958–9 edition) shows a low streamlined fin strikingly reminiscent of the Type XXI, with twin 25mm gun turrets superimposed at the after end. The AA guns appear to have been removed in the late 1950s, as photographs taken from 1960 onwards show the fin 'cleaned up' at its after end. Shortly afterwards, the fin appears to have been further modified, with a prominent snorkel housing being added topsides. This housing was later to be further extended, with the snorkel exhaust protruding from its after edge, giving the 'Zulu' its final configuration (generally known as 'Zulu IV').

Opposite page: *The configuration of the original 'Zulu' fin can be clearly made out in this view of a 'Zulu IV'. The additional section above it accommodates most of the sensor masts and terminates in a curved 'lip' housing the diesel exhaust. (MoD, 1981)*

The 'Quebec'

The small submarine designed for coastal defence was well-established in the Soviet naval inventory, and at the end of the Second World War a number of units of the 'M' class were still under construction at the Sudomekh Shipyard, Leningrad. They were subsequently launched and fitted out, and remained in service until disposed of in the 1960s.

These boats, which belonged to Series XV, were of superior design and construction to the early 'M'-class boats. They were significantly larger, and had two-shaft propulsion machinery and four torpedo tubes (the early boats had only two). Their completion postwar proves that the concept of the small coastal submarine had not been discredited as a result of war experience. On the contrary, the acquisition of German technology was to provide the necessary impetus for a new design, the 'Quebec'.

As a part of their allocation of war spoils, the Soviet Navy had received a coastal submarine of the German Type XXIII class. The Type XXIII was similar in conception to the Soviet 'Malutka' in that it was designed for prefabricated construction in non-specialist shipyards, and was transportable by railway between the various maritime theatres. In technological terms it was a scaled-down version of the Type XXI, and like its big brother had been developed from a Walter design, with conventional diesel-electric propulsion substituted for the closed-cycle turbine of the latter.

Soviet interest in a small coastal boat employing closed-cycle propulsion dated from before the War, when a 'Special Purpose Regenerating Power Unit' (*Regenerativnyi Yedinyi Dvigatiel Osobovo Naznacheniya*) had been installed experimentally aboard two 'M'-class boats of the Series XIIbis. The experiments were subsequently abandoned, but Soviet interest was revived postwar. The installation of closed-cycle propulsion aboard small coastal submarines, which were essentially short-range 'interceptors', was clearly a relatively 'low-risk' proposition compared with the development of a viable closed-cycle propulsion system for larger boats designed for extended patrols.

The first submarine of the 'Quebec' class was probably laid down in 1952 or 1953 at the Sudomekh

Below: *The small size of the 'Quebec' is readily apparent in this photograph taken at a Leningrad Navy Day during the 1960s. Behind the submarine is a destroyer of the* Skory *class. The other vessels are minesweepers. Note the light-grey paintwork which was a feature of Soviet submarines during the early post-war period. (TASS)*

Shipyard, Leningrad, and was completed in 1954. With a surfaced displacement of around 400 tons, the 'Quebec' was significantly larger than both the later Soviet 'M' series and the German Type XXIII. This increase in size represented an attempt to come to terms with the stability problems experienced with the warbuilt types, but was also related to the hybrid propulsion system adopted, which consumed more space than the conventional diesel installation of the latter.

The German Type XXIII had only a single shaft, as had the prewar Soviet 'Malutkas'. Twin-shaft propulsion was subsequently adopted for the Series XV boats of the 'M' class. The 'Quebec', on the other hand, was initially fitted with three shafts, the outer two shafts being powered by conventional diesels and electric motors with the centre shaft driven by a Kreislauf closed-cycle diesel, which could be operated when the boat was running submerged.

Serious problems were apparently experienced with the closed-cycle diesel, and there were a number of accidents which caused injury to personnel. It was therefore decided to abandon the new propulsion unit and to instal conventional diesels and motors on all three shafts. The maximum underwater speed produced by the revised installation has been estimated at 16 knots – double that of the 'M' Series XV, and four knots faster than the single-shaft German Type XXIII.

The increase in size over earlier types enabled the Soviet designers to make improvements both in endurance and in armament. In addition to the torpedoes carried in the four full-size 533mm bow tubes, four reloads could be accommodated. The 'Quebecs' were also fitted with the full range of sensors available to contemporary Soviet submarines.

The 'Quebec', unlike other postwar Soviet submarine classes, appears to have been a single-hull design with a flat deck-casing. The tall fin, which had a prominent raised housing for a snorkel at its after end, was clearly designed to facilitate surface operation in rough weather and to enable the submarine to operate comfortably at snorkel depth. Early boats were apparently fitted with a twin 25mm AA mounting forward of the fin, but this was quickly removed as it was found to have an adverse effect on underwater performance.

As with the 'Whiskey' and 'Zulu' classes, there was considerable initial confusion in the West regarding the size, purpose and number of these boats. The first mention of the 'Quebec' in *Jane's* was in the 1958–9 edition, where the submarine was described as an 'improved Shch' and credited with a surfaced displacement of 750 tons! The number in service was estimated at 40, rising to 50 the following year. Later estimates were to vary between 20 and 30, the latter figure being the one generally accepted by current reference sources.

There is some evidence to suggest that these 30 boats, which were deployed exclusively in the Baltic and the Black Sea fleet areas, constituted only the first instalment in a much larger programme intended to run on into the early 1960s. Michael McCGwire[5] has estimated a planned production run of at least 96 boats. Problems with the propulsion system undoubtedly caused delays in the programme and prevented the Sudomekh Shipyard from achieving mass production within the time-scale initially envisaged. The premature curtailment of the programme, however, is almost certainly more closely related to fundamental changes in Soviet Naval policy during the mid-1950s, which also brought about the termination of the construction of the 'Whiskey' successor, the 'Romeo'.

Sensor equipment

Soviet interest in hydroacoustic detection and underwater communications dates from well before the

'QUEBEC' CLASS

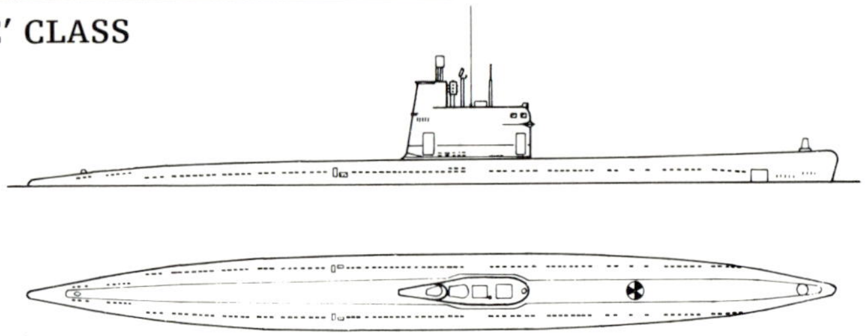

'QUEBEC' CLASS

Soviet designation: Podvodnaya Lodka (PL) (Submarine)

Construction:
Built at Sudomekh Shipyard, Leningrad. 30 units completed 1954–7: three in 1954, ten in 1955, twelve in 1956, five in 1957.

Displacement:
Surfaced: 400 tons
Submerged: 540 tons

Dimensions:
Length: 56m (185ft)
Beam: 5.1m (17ft)
Draught: 3.8m (13ft)

Propulsion:
Diesel/electric drive on three shafts (see text); three diesels, 3,000bhp for 18 knots surfaced; three electric motors, 2,200hp for 16 knots submerged.

Complement 30

Weapons:
Four bow tubes of 533mm (21in) diameter for anti-ship torpedoes; maximum capacity eight torpedoes.

Sensors:
HF passive sonar; Tamir 5 HF active sonar in some; search and attack periscopes; Snoop Plate surface surveillance radar; Stop Light ESM; Quad Loop DF.

Communications:
HF and VHF radio.

Above: *Although the caption for this Soviet photograph states that the model being presented to Captain S. Lysov is of his wartime submarine command, K21, the model clearly depicts a submarine of the 'Zulu' class. The position of the bow and stern torpedo tubes and the layout of the after control surfaces appear to be accurate. Note the large bulb projecting below the bow section. It probably housed a passive acoustic array modelled on the* Balkon Gerät *of the German Type XXI. (Novosti)*

outbreak of the Second World War. The Underwater Sound Laboratory was established in Leningrad in 1930, and a Navy Department of Hydroacoustics was set up shortly afterwards. A Naval Science Research Communications Institute was also in operation by 1934.

The earliest Soviet sonar appears to have been a passive high-frequency model comprising two magnetostrictive directional-effect receivers in an aluminium housing. Production models were subsequently developed, and the Soviets claim to have had Mars passive sonars aboard no fewer than 176 of their submarines prior to the German invasion in July 1941. A new system, designated Drakon-129, was installed aboard five submarines in 1942, and improvements were made in 1944. However, the detection equipment available to the Soviets during the war appears to have been relatively primitive and ineffectual.

From 1942 onwards, the Soviet Navy began to receive active acoustic detection sets of British or US origin. These, however, were installed in active fighting units (generally destroyers) as soon as they arrived, and it would be some years before the Soviets had either the time or the opportunity to examine and evaluate this equipment under peacetime conditions.

German submarine technology was also accessible from 1945, but it was to be a further two years before the fully fitted German Type XXIs, with their large bulbous chin domes housing the Balkon Gerät passive array, would be available for trials and testing.

These outside influences took time to absorb, and some of the assumptions commonly made in reference books about the sonar equipment installed in the early Soviet post-war classes are difficult to substantiate. There is, for example, no evidence of the mechanically rotated active sonar transducers generally fitted in Western post-war submarines. In virtually all the Western boats these were housed in a prominent sonar dome above the bow. Similar equipment was fitted in Soviet submarines only from 1958, when a new generation led by the 'Foxtrot', 'Golf' and 'Romeo' classes began to appear. The first photographs of submarines of the 'Whiskey', 'Zulu' and 'Quebec' classes with bow sonar domes date from the early 1960s, and many of the early boats were retro-fitted with this equipment only during the second half of that decade.

The apparent lack of Soviet interest in active sonar during the early post-war years is undoubtedly related to the relatively low priority accorded to anti-submarine warfare during the period. Active (as opposed to passive) sonar was most useful for providing accurate fire control data when hunting other submarines. This was a mission being accorded increasing priority in the West because of the perceived threat presented by the large Soviet submarine fleet. The Soviets, on the other hand, were concerned only with surface targets, and preferred passive hydroacoustic detection methods, which were adequate to determine bearing and did not serve as a beacon for hostile anti-submarine units.

It is possible that the early post-war submarines built for the Soviet Navy were fitted only with chin-mounted passive arrays developed from wartime German models, plus an underwater telephone derived from the wartime Sirius equipment.

So few photographs have been published of 'Whiskeys' and 'Zulus' in their early configurations that it is difficult to determine which mast-mounted sensors were initially fitted. The 'Whiskey' appears to have been fitted with search and attack periscopes, a whip aerial for HF communications, and a Quad Loop direction-finder mounted on a retractable mast. It is not clear whether the early boats carried surveillance radar, and it seems unlikely that ESM equipment to detect hostile radar emissions could have been developed in time for installation in the first units com-

372

pleted. This theory is supported by the subsequent displacement of the Quad Loop DF aerial to accommodate the new ESM mast when the latter became available.

Later boats appear to have had a full outfit of mast-mounted sensors in an extensively modified fin, and all of the early boats were subsequently retro-fitted to the same standard. In the 'Whiskey V', as the definitive variant is generally known, there were search and attack periscopes, a snorkel, a Snoop Plate surface surveillance radar, and a Stop Light ESM antenna, all of which were mounted on retractable masts. The Quad Loop DF aerial was re-located in a fixed position on the forward edge of the snorkel exhaust, which was angled out from the after edge of the fin. In addition to the whip aerial for HF communications there was a shorter whip for VHF/UHF tactical communications located at the top of a retractable mast.

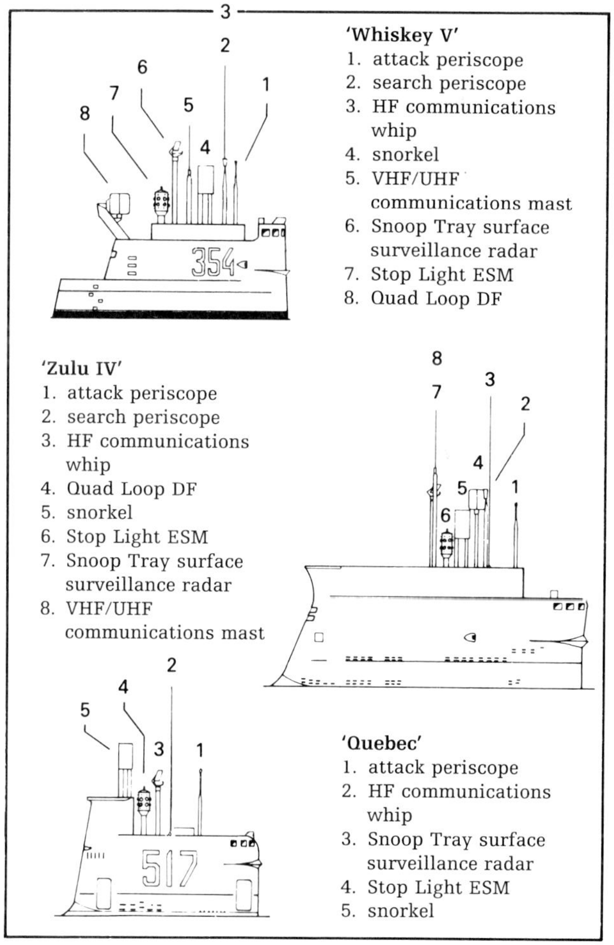

Left: *'Whiskey' 372 in the English Channel during the late 1960s. By this time all 'Whiskeys' were fitted with the small dome for a Tamir 5L sonar atop the bow. (Skyfotos, 1967)*

The difficulty of accommodating the full array of sensors plus a snorkel in a relatively small fin is illustrated by the problem experienced in re-locating the Quad Loop aerial. The 'Zulu', being a larger submarine with an exceptionally long fin, proved more susceptible to modification, and all sensors including Quad Loop were mounted on retractable masts.

The 'Quebec' benefited from being laid down some three years after the 'Zulu', and was fitted with a fin of more modern design which influenced that of later boats. The snorkel, togehter with its exhaust, was housed in a raised section at the after end of the fin. Published photographs show only one periscope in the raised position, so it may be that the functions of search and attack were combined in a single periscope, as in post-war Swedish submarines. A Snoop Plate surveillance radar and Stop Light ESM were, however, incorporated from the outset, and it seems logical to conclude that a mast-mounted Quad Loop DF aerial was also fitted.

The above-water sensor outfits of the Soviet post-war submarines therefore correspond fairly closely to those of Western boats of the period, but many of the items of equipment detailed above were available only from the mid-1950s, placing the Soviet Navy 5–10 years behind their Western counterparts in technological terms.

There were also differences of emphasis which relate to Soviet tactical doctrine. There was a preference for passive detection and surveillance systems over active systems. In part this reflected Soviet thinking about the essentially covert nature of submarine operations. However, it is also important to remember that in the Soviet concept of naval operations the submarine would not be working alone, but would in effect be a weapons carrier supported by an extensive surveillance network which would include reconnaissance aircraft equipped with active radar, and quite possibly other submarines in visual contact with the enemy force.

The sensors carried by Soviet submarines of the post-war generation therefore had the following functions: they provided for the initial detection of hostile surface forces; they could be used to confirm a bearing passed to the submarine by the control post ashore, enabling the submarine to take up an attacking position ahead of the enemy force; and they provided the submarine with a means of detecting the presence of hostile aircraft when operating on the surface. The installation of ESM equipment capable of detecting aircraft radar emissions appears to have been accompanied by the removal of all AA guns from Soviet submarines.

Because the submarine operated as part of a larger force, and not as a 'lone wolf' in the Western mould, good communications were crucial to the success of the operation. The Soviets therefore stressed not only long-range (HF) communications with the control post, but also short-range (VHF/UHF) 'tactical' communications to enable the submarine to collaborate with other units

Below: *Close-up of the fin of a 'Whiskey V'. Besides the HF communications whip aerial, the VHF/VHF communications mast and the Stop Light ESM antenna can be seen in the raised position. The Quad Loop DF antenna is mounted on the forward side of the diesel exhaust. The head of the snorkel can be made out between the HF whip and the VHF radio mast. Note the 'bridge windows' which are a feature of all Soviet submarines, and the spray deflector beneath. (Skyfotos, 1968)*

Right: *A 'Zulu' running at speed on the surface. Note the raised snorkel, behind which can be seen the Stop Light ESM antenna and the VHF/UHF tactical communications mast. (MoD, 1970)*

in its operational area. An underwater telephone performed a similar function when the boat was submerged.

Weapons

Some of the early 'Whiskeys', and possibly some 'Zulus', were initially fitted with a deck gun. Most sources agree that this was of 100mm calibre, although some sources have talked of single 76mm and twin 57mm mountings. In the absence of photographic evidence the 100mm calibre seems the most logical, given that this was the standard calibre for the Soviet warbuilt submarines.

Soviet pre-war and warbuilt submarines also carried one or two single 45mm anti-aircraft guns. This model was replaced post-war by a new 25mm/60cal twin gun mounting which, in the submarines, was installed in an enclosed turret housing. These gun housings were removed in the mid/late-1950s. The twin 25mm mounting, which featured superimposed gun barrels, was essentially the same model fitted in large numbers of small surface ships built for the Soviet Navy and for the navies of the Warsaw Pact countries during the post-war period, and had a rate of fire estimated at 150–200rpm per barrel.

The torpedoes in service with the Soviet Navy during the war years had proved unreliable. The Soviets therefore developed a new model, generally designated Type 53-VA in the West, which would equip the submarines of the post-war generation. The Type 53-VA was a 533mm heavyweight torpedo similar in many respects to the British Mk 8 and the US Mk 14, with a length of approximately 7.75m and conventional steam propulsion. The guidance system is reported to have been modelled on the FAT-II pattern-running device incorporated in the German Type G7 torpedo developed during the Second World War. Detonation was by impact or magnetic fuzing. The earliest variant of the Type 53-VA is thought to have entered service in 1951, the same year which saw the completion of the first units of the 'Whiskey' class.

At least one source[6] states that this torpedo was superseded by a variant with HTP (High Test Peroxide) propulsion in about 1956. If this is the case, Soviet torpedo development would have followed the same pattern as developments in the Royal Navy and US Navy. The US Navy's Mk 16, powered by Navol (hydrogen peroxide), proved relatively successful, and remained in service until 1975. However, the British HTP-powered Mk 12, code-named 'Fancy', was abandoned following a propellant explosion which sank the submarine *Sidon* in 1955. It seems logical to assume that the Soviets would have experimented with HTP propulsion for their torpedoes, but with how much success must remain open to question. Given Soviet preference for reliable, robust, easily maintainable equipment over more sophisticated but complex and 'fragile' systems, it is unlikely that such a torpedo would have completely superseded its conventionally powered predecessor, which has probably outlasted it in much the same way that the US Mk 14 torpedo continued in service after the Navol-powered Mk 16 was withdrawn.

2. The Nuclear Age

Joseph Stalin died in March 1953. He had been in power for 30 years, a period that had seen the industrial regeneration of the Soviet Union and its emergence as a military superpower. It was also a period marked by brutal repression and the elimination of all opposition both within the Soviet Union and in those countries of eastern Europe that had fallen under Soviet political domination. It was therefore inevitable that Stalin's death should result in a power struggle.

Stalin was initially succeeded by Georgi Malenkov, who proceeded to institute a major review of defence policy in 1954. It had become clear that traditional Soviet military strategy, which had been based on defence in depth employing large numbers of tanks, aircraft, submarines and torpedo-boats, would have to be modified in the light of the development of nuclear weapons. Unlike conventional explosives, nuclear weapons could wreak devastation over a wide area, and were therefore most effective against the tight concentrations of forces favoured by the Soviets.

Moreover, the atomic bomb was no longer exclusively the property of the Americans and the British. The Soviets exploded their first hydrogen bomb in August 1953, five months after Stalin's death and less than one year after the United States had first tested their own device. Now that the Soviet Union had its own nuclear capability, Soviet military strategy would have to change in order to accommodate it. There would have to be a major restructuring of the armed forces, and fundamental changes in tactical thinking. Generals and admirals who owed their positions to the tactical successes of the Great Patriotic War were naturally less than enthusiastic about many of these changes, which would be resisted just as strongly in the Soviet Union as they were in the West during the early post-war years.

Left: *The advent of the nuclear-powered submarine, in the shape of the US Navy's* Nautilus, *opened up a new dimension in underwater operations.* Nautilus *was to provide the stimulus for a large Soviet nuclear submarine construction programme, beginning with the 'November'. (US Navy, 1965)*

After only two years in office, which were marked by serious policy disagreements within the Politburo, Malenkov was overthrown and replaced by Nikolai Krushchev, who became First Secretary of the Communist Party in February 1955. Krushchev was a forceful individual with a sound war record who immediately made his mark on Soviet defence policy. Within a year the naval C-in-C, Admiral Kuznetsov, had been dismissed and replaced by a younger man, Admiral Sergei Gorshkov.

Kuznetsov had been appointed by Stalin in 1951. He was a 'big ship' sailor in the traditional mould, and was closely identified with Stalin's dream of a large ocean-going surface fleet comprising cruisers, battlecruisers and even aircraft carriers. When Krushchev attacked the naval construction programme as expensive and irrelevant, Kuznetsov defended it, and paid with his job.

Gorshkov's appointment as C-in-C of the Navy at the relatively tender age of 45 marked him out as a forceful and ambitious. His brief was to preside over a revolution in Soviet naval affairs which would involve a fundamental reappraisal of Soviet naval missions and of the types of vessel best suited to perform them.

Krushchev believed that the naval scene had been transformed by the advent of nuclear weapons. Allied to the rocket technology that was now becoming available, nuclear weapons would enable individual naval units to attack and disable entire task forces from beyond the radar horizon. Conventional surface fleets, including those based around aircraft carriers, had been rendered obsolete at a stroke. Moreover, there was no longer a requirement for large numbers of low-value units to defend the outer zone of Soviet sea-space. Rather, what was needed was a smaller number of more capable platforms, large enough to accommodate and launch nuclear missiles against hostile surface forces.

Krushchev was also enthusiastic about the potential of nuclear propulsion for submarines. He was angry that the Soviet Navy had been so slow to see this potential, and that it had wasted so much effort in pursuing other approaches to closed-cycle propulsion. In the event, neither the British nor the Russians were able to develop an operational Walter turbine, whereas the US Navy had its first nuclear-powered submarine, *Nautilus*, in service by late 1954.

The primary reason for the failure of both the British and the Russians to focus on the potential benefits of nuclear submarine propulsion was the respect, amounting almost to reverence, with which they tended to regard German submarine technology. Krushchev, for all his invective against the United States in his public appearances, had a tremendous admiration for the American ability to analyse problems logically, without being stymied by established military thinking, and to harness technology to the production of equipment which could best perform US missions.

It is also of interest to note that under Krushchev's protégé, Gorshkov, the missions and structure of the Soviet Navy would increasingly be tailored to the maritime threat presented by the US Navy, rather than to the more general threat presented by the navies of the Western Alliance. Changes in US naval strategy and tactical thinking would henceforth provide the major impetus for changes of naval policy in the Soviet Union, and for the development of new technologies to counter specific US naval threats.

The new generation of submarines

By the time the post-Stalin defence debate had ended, the designs for the successors to the 'Whiskey' and the 'Zulu' were already on the drawing-board. The 'Quebec', which had begun construction only in 1953–4, was to have continued in production probably into the early 1960s.

The 'Whiskey' successor was the 'Romeo', the first of which was laid down about 1956, probably at the Krasnaya Sormovo Shipyard, Gorky. At about the same time the lead boat of the 'Foxtrot' class, which would supersede the 'Zulu', was laid down at the Sudomekh Shipyard, Leningrad. Both were intended for series production, with the numerical ratio between the medium 'Romeo'-class boats and the ocean-going 'Foxtrot' being similar to that established by their predecessors.

There was little that was radically different in the design of either boat. The hull-form of the 'Romeo' was more streamlined than that of the 'Whiskey', the prominent saddle tanks of the latter being abandoned in favour of a cigar-shaped outer hull casing. However, dimensions, internal layout and propulsion machinery were virtually identical, as were the control surfaces, which comprised retractable bow planes, stern planes mounted on horizontal stabilizers abaft the twin propellers, and a single large rudder integrated into the stern.

Some incremental improvements were made. There was an additional pair of torpedo tubes in the bow, making a total of six tubes forward and two aft. Standard reference sources also state that more powerful electric motors were fitted, although there appears to have been no corresponding increase in underwater speed.

There were even fewer differences between the 'Zulu' and the 'Foxtrot', which was essentially a repeat design. In terms of overall dimensions, displacement,

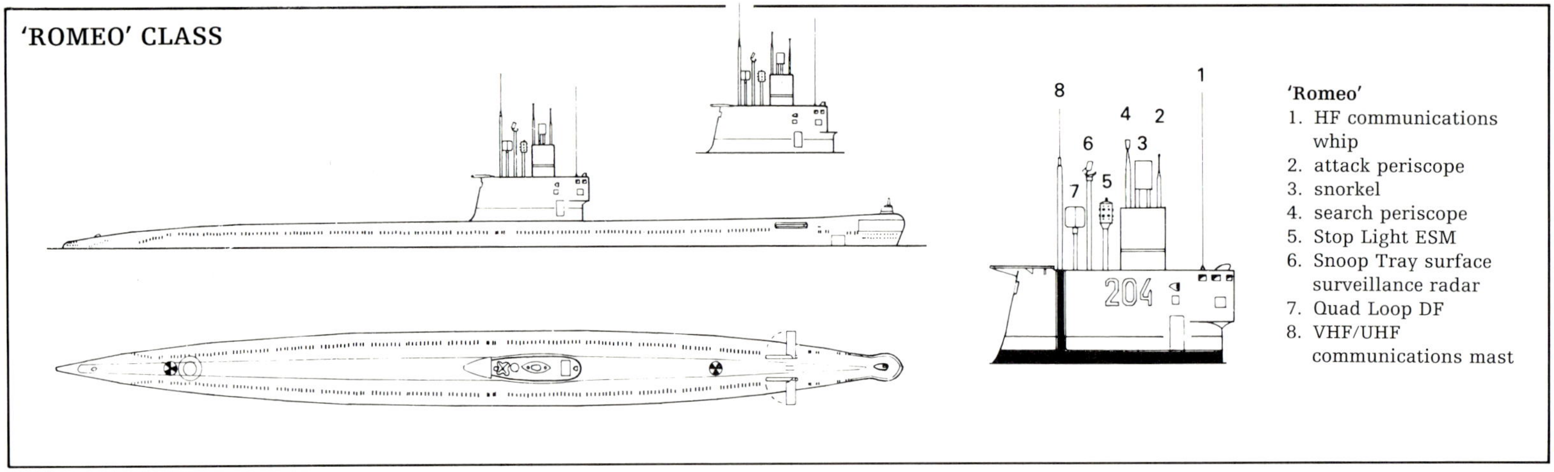

hull-form, internal layout, propulsion and armament the two types were virtually identical.

The principal distinguishing feature between the new submarines and the first-generation boats was the sensor outfit, which was altogether more advanced and elaborate than that of their predecessors. Both the 'Romeo' and the 'Foxtrot' had a remodelled bow to accommodate new models of active and passive sonar, and both were designed from the outset to carry a full range of mast-mounted sensors in a remodelled fin.

In both cases the upper section of the bow was given a broadly circular configuration in order to accommodate a Feniks passive array. The array itself appears to comprise a crown of hydrophones, probably in triple staves, approximately two metres in diameter. Atop the bow is a small streamlined radome of elliptical cross-section no more than half a metre across. This is reported to house the Herkules active sonar, a high-frequency mechanically scanned 'pinger' used to provide precise ranging on a target. A small cone atop the

Above left: *The inset shows a later modification to the conning tower.*

Left: *The 'Romeo' was designed as a successor to the 'Whiskey'. Overall dimensions were similar, but the hull-form was closer to that of the larger 'Zulu' and a new outfit of sonars was installed. (US Navy, 1973)*

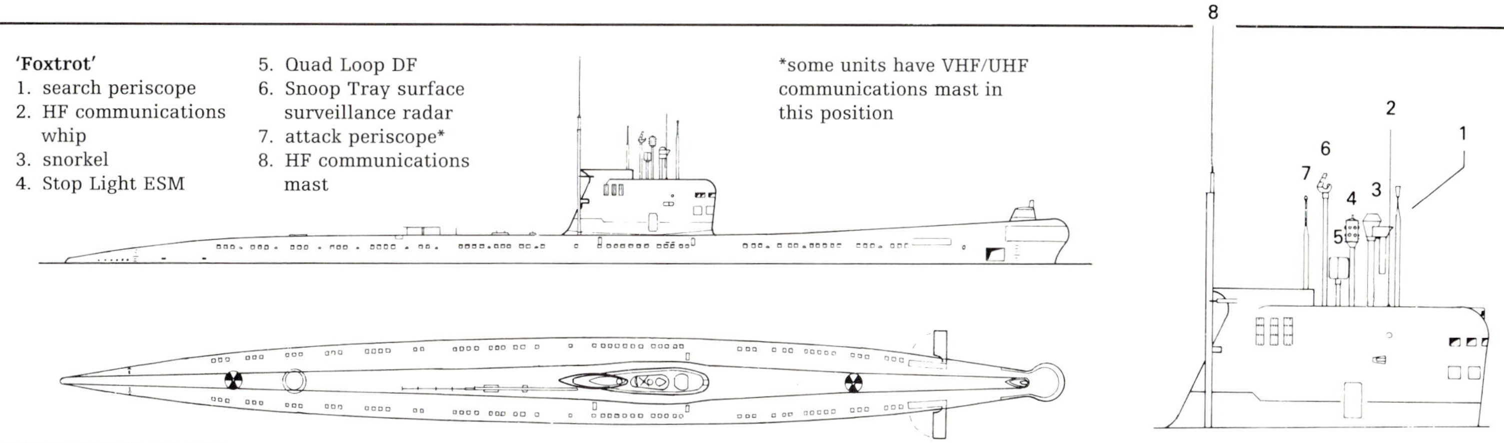

'FOXTROT' CLASS

radome is thought to house the antenna for an underwater telephone, probably derived from the wartime Sirius.

The first 'Romeos' and 'Foxtrots' completed appear from early photographs to have had a light-grey paint-scheme. Later photographs, however, show both types in the standard black colour-scheme but with large unpainted 'windows' around both the Feniks and the Herkules sonars. This suggests that the black paint employed had sonar-absorbent qualities which would have adversely affected the performance of the boats' own sonar equipment.

In addition to the prominent sonar 'windows' wrapped around the bows, the 'Foxtrot' had four panels approximately one metre square located around the forward edge of the fin. The purpose of this equipment is unclear, although it appears reasonable to assume that it is sonar-related. It is possible that these panels conceal spot hydrophones which together form a passive fire-control system not unlike the US Navy's

Right: *Big brother and little sister. This 1983 view of a 'Whiskey' lying outside a 'Foxtrot' was taken at a Black Sea naval base. The disparity in size is immediately apparent, as is the more extensive sonar outfit of the 'Foxtrot'. The prominent unpainted 'window' wrapped around the bow conceals a Feniks medium-frequency passive array; the transducer for the Herkules active sonar is housed in a streamlined dome atop the bow. The cone above it is thought to house an underwater telephone. Note the difference in size between the Herkules sonar dome of the 'Foxtrot' and the Tamir 5 sonar dome of the 'Whiskey'. (TASS)*

PUFFS. The latter system comprises three or four hydrophones mounted in the same plane on the sides of the hull, the target's range and bearing being calculated by measuring the time delay between the reception of a signal at each of the hydrophones. The Soviet system is almost certainly less sophisticated; by employing hydrophones close together at fixed angles to one another it is presumably possible to obtain good bearing definition, but the active 'pinger' would probably have to be used for a range solution adequate for weapons launch.

The numerous modifications made to the fins of the original 'Whiskey'- and 'Zulu'-class boats have been detailed in the previous chapter. The removal of all guns and the development of the snorkel and electronic surveillance equipment necessitated major reconstruction. The fins of the new boats were therefore completely redesigned to accommodate all these systems.

The fin of the 'Foxtrot' was closely modelled on that of the smaller 'Quebec'. It had a rounded forward section incorporating an enclosed bridge, and the after part, which housed the snorkel exhaust, was raised. The snorkel mast itself was forward of the raised section, with the search and attack periscopes forward and the masts carrying the electronic surveillance sensors grouped abaft it. The VHF/UHF (tactical) communications mast was housed in the forward end of the raised section. For long-range, high-frequency communications there was a fixed whip aerial atop the fin, and a second aerial mounted atop a prominent hinged mast, which could be folded back into a recess

'ROMEO' CLASS

Soviet designation: Podvodnaya Lodka (PL) (Submarine)

Construction:
Built at Krasnaya Sormovo Shipyard, Gorky.
Twenty units completed 1958–62: two in 1958, three in 1959, five in 1960, nine in 1961, one in 1962.

Displacement:
Surfaced: 1,300 tons
Submerged: 1,700 tons

Dimensions:
Length: 77m (253ft)
Beam: 6.7m (22ft)
Draught: 4.9m (16ft)

Propulsion:
Diesel/electric drive on two shafts; two Type 37D diesels, 4,000bhp for 15.5 knots surfaced; two electric motors, 2,700hp for 13 knots submerged.

Complement 55

Weapons:
Six bow tubes, two stern tubes, all of 533mm (21in) diameter, for anti-ship torpedoes; maximum capacity fourteen torpedoes.

Sensors:
HF passive sonar; Herkules HF active sonar; search and attack periscopes; Snoop Plate surface surveillance radar; Stop Light ESM; Quad Loop DF.

Communications:
HF and VHF radio.

Above: *A 'Foxtrot' running at speed on the surface in the North Atlantic. (US Navy, 1979)*

'FOXTROT' CLASS

Soviet designation: Podvodnaya Lodka (PL) (Submarine)

Construction:
Built at Sudomekh Shipyard, Leningrad; Shipyard No. 402, Severodvinsk.
62 units completed 1958–71.

Displacement:
Surfaced: 1,950 tons
Submerged: 2,400 tons

Dimensions:
Length: 91.5m (300ft)
Beam: 7.5m (25ft)
Draught: 6m (20ft)

Propulsion:
Diesel/electric drive on three shafts; three diesels, 6,000bhp for 16 knots surfaced; three electric motors, 5,300hp for 15.5 knots submerged.

Complement 78

Weapons:
Six bow tubes, four stern tubes, all of 533mm (21in) diameter, for anti-ship/anti-submarine torpedoes; maximum capacity twenty-two torpedoes.

Sensors:
Feniks MF passive sonar; Herkules HF active sonar; fin-mounted passive FC arrays; search and attack periscopes; Snoop Tray surface surveillance radar; Stop Light ESM; Quad Loop DF.

Communications:
HF and VHF radio.

Above: *A 'Romeo' in the English Channel in 1972. Note the distinctive elliptical 'sleeve' projecting from the conning tower. (Skyfotos)*

in the after deck casing when not in use. The emphasis on long-range communications is an indication of the importance accorded to the scouting mission of these submarines, and of the distance from their Soviet bases at which it was intended that they should operate.

The fin of the 'Romeo' was of altogether different design. It was of similar height to the fin of the later 'Whiskeys', but there was no 'step' at the after end, the upper section being extended aft to create sufficient space for all the necessary sensor masts. At the forward end there was an enclosed bridge which projected over the forward deck casing. This configuration appears to have been adopted to enable the designers to dispense with the spray deflectors which had been fitted in earlier types. Below the bridge windows there were square panels to port and to starboard which may have been a smaller version of the passive fire control system which equips the 'Foxtrot'.

The snorkel and the two periscopes were accommodated in a distinctive elliptical housing projecting some three metres above the top of the fin. The purpose of this feature was to enable the submarine to operate at greater depth when snorkelling or attacking. Its adoption suggests that problems may have been experienced with the ingestion of water, possibly due to difficulties in maintaining depth control in heavy seas, when snorkels were first fitted in the 'Whiskey'. The electronic surveillance sensors and retractable VHF/UHF mast were grouped together abaft the snorkel housing, and there was only a single whip aerial for HF communications, these boats being intended for less distant patrols in defence of Soviet sea-space.

A change of direction

The nuclear age ushered in major changes of strategy for both sides. Krushchev was particularly conscious that massed conventional forces could no longer provide a protective buffer around Soviet territory. There were equally strong doubts in the United States regarding the future viability of large amphibious operations. In the immediate post-war period, even the supremacy of the aircraft carrier, which had become the cornerstone of US naval strategy during the Pacific War, appeared to be threatened by the awesome power of nuclear weapons.

However, once the experience of the Korean War had re-established the aircraft carrier as a key element in US maritime strategy, the US Navy took the opportunity to secure its position by adapting the carriers to a nuclear strike role. A new generation of large 'supercarriers' was sanctioned, designed to operate nuclear-capable, jet-propelled bombers with an effective strike radius in excess of 1,000 nautical miles.

These developments effectively outranged Soviet countermeasures, which were founded on theatre-based concentric defence zones, the outer boundaries of which were to have been patrolled by medium submarines of the 'Whiskey' and 'Romeo' classes. In future, Soviet submarines could not afford to lie in wait for the approaching task forces, but would have to seek out – and it was to be hoped destroy – the carriers before they reached striking range.

Medium submarines armed with conventional torpedoes were inadequate for the task. The 'Romeo' programme was therefore drastically curtailed. Of the 570 units reported to have been initially projected[1] only twenty were completed, construction being terminated in 1962. A number of these boats were transferred to other allied or friendly countries during the 1960s and 1970s. They were still useful for operations in enclosed and shallow seas such as the Baltic and the Black Sea, but the Soviets clearly regarded them as being of limited value for the ocean operations which would play an increasingly important part in Soviet naval policy.

Construction of the larger 'Foxtrot' class would continue into the late 1960s, and appears to have been relatively unaffected by these changes in Soviet defensive strategy. The 'Foxtrots' would also be an effective platform for a new variant of the Type 53-VA torpedo, armed with a nuclear warhead.

To complement the 'Foxtrot', however, a new class of submarine powered by nuclear reactors was rushed

Left: *A 'Foxtrot' photographed by an RAF Nimrod south of the Faroes in 1974. Note the raised snorkel mast, with the Snoop Tray surveillance radar just visible behind it.*

Right: *This unit of the 'Foxtrot' class has the attack periscope in the position usually occupied by the VHF/UHF radio mast, at the forward end of the raised section of the conning tower. The snorkel mast is also in the raised position, and the Quad Loop DF antenna is visible just behind it. (MoD, 1988)*

Right: *During the 1970s at least one unit of the 'Romeo' class had an extension housing a towing hawser fitted at the base of the conning tower. (US Navy, 1986)*

Right: *A 'Foxtrot' in the English Channel in 1967. It was at this time that regular deployments from the Northern Fleet to the Mediterranean began. (Skyfotos)*

Left: *Two submarines of the 'Foxtrot' class in northern waters. By the mid-1970s 31 boats out of an estimated 60 in the class were in service with the Soviet Northern Fleet. (TASS, 1983)*

Above: *Close-up of the conning tower of a 'Foxtrot'. The snorkel mast is in the raised position, as are the masts for the Snoop Tray surveillance radar and for VHF/UHF tactical radio communications. The tall hinged HF communications mast at the after end of the conning tower can be lowered into a recess in the after deck casing. (MoD, 1975)*

into service. The first unit of the 'November' class was commissioned in August 1958. This in itself was a remarkable achievement; according to Soviet sources development of a submarine reactor began only in 1953, and there are indications that the programme did not receive particularly high priority until Krushchev came to power.

A comparison with the development of the US Navy's early nuclear submarines is revealing. The Americans built two prototypes, *Nautilus* and *Seawolf* (completed in 1954 and 1957 respectively), with competitive reactors, plus a small production class (the *Skate* class) with a conventional hull based on *Nautilus*. The first US attack submarine to combine a nuclear reactor with the revolutionary 'Albacore' hull-form was the *Skipjack*, which was laid down in 1956 and completed in 1959. Parallel development of prototypes continued, and there was a two-year gap between the completion of *Skipjack* and the second unit of the class.

The Soviet Navy, on the other hand, bypassed the prototype stage altogether and launched straight into series production of the 'November', which featured a modern 'tear-drop' hull-form as well as nuclear propulsion. The compressed development process provides a remarkable testimony to the forcefulness of Krushchev's personality. The degree of urgency which he brought to the nuclear submarine programme resulted in an apparent turn-around in the technological imbalance between the US Navy and the Soviet Navy. To the outside observer, it would seem that the Soviet Navy had started the race four years behind the Americans, and had finished one year ahead.

Construction dates do not tell the full story, however. There can be little doubt that the Soviet nuclear submarine programme was a risky enterprise, in which all the technological eggs were being placed into a single untried basket. The decision to dispense with the prototype stage and to proceed to full series production meant that there was no opportunity to monitor the nuclear machinery in operation, or to correct design flaws that might have serious consequences for the safety of the submarine and its crew.

The early Soviet nuclear boats were to have more than their fair share of accidents during the next two decades. Some were simply machinery breakdowns which were to result in the submarine being towed home to its base. Others, however, were more serious, with fires and radiation leaks causing many casualties among crews. For all its claims to represent the latest in 'high technology', the 'November' was essentially a hurriedly designed and hurriedly assembled prototype which, because of the political priority accorded to the nuclear submarine programme, was rushed into series production. Moreover, even before the 'November' had attained operational status, two other types of Soviet nuclear submarine had been designed around its leaky and unreliable propulsion system.

The design

The 'November' was a large submarine even by Soviet standards. With a length of 110 metres and a submerged displacement of 5,300 tons, it dwarfed the US Navy's *Skipjack* (76.7m and 3,500 tons respectively). It established a pattern for Soviet nuclear submarine construction which would be continued over the next three decades, with the notable exception of the 'Alfa'.

The US Navy had been quick to see that the advent of nuclear reactors would favour single-hull construction. Conventionally powered submarines had to carry large quantities of oil fuel in tanks which were generally located outside the pressure hull. Double-hull construction, in which the space between the pressure hull and the outer casing could be utilized for fuel stowage, ballast tanks and blowing systems, was therefore favoured by the majority of submarine designers.

Left: *Whereas a conventional hull-form was retained by the US Navy for its early nuclear-powered submarines, the Soviet designers opted from the outset for an advanced, 'tear-drop' hull which was combined with a low, streamlined fin for optimum underwater performance. (US Navy)*

One consequence of nuclear power was that the requirement for fuel stowage was eliminated. Ballast tanks would of course still be required, but by locating these at either end of the submarine they could be accommodated within a single hull. Other systems previously carried between the inner and outer 'skins' could be housed within the pressure cylinder itself.

There were a number of important advantages to be gained by moving to single-hull construction. The weight of the hull structure was considerably reduced. Moreover, for a pressure hull of a given volume there would be a significant reduction in the overall dimensions of the submarine. Since the horsepower required to propel a submarine at a given speed is dictated by wetted surface area, a single-hull design would have superior underwater speed for a fixed power plant. Alternatively, it could be propelled at the same speed as a comparable double-hulled design by a propulsion plant with a lower power rating. This would open up the design option of smaller size and higher speed, or greater internal hull volume which could be used for weapons stowage, sophisticated electronics, access to machinery, and superior habitability (a feature increasingly necessary with the advent of long underwater patrols).

The US Navy adopted single-hull construction for its nuclear-powered submarines from the outset. Even *Nautilus*, with a relatively conventional hull-form, was a single-hull design. However, the advantages of single-hull construction were fully realized only with the *Skipjack* class, which combined relatively small dimensions and a modern, 'Albacore' hull-form with a maximum speed in excess of 30 knots.

The 'November', which in accordance with traditional Soviet practice featured a double hull, could not approach this sort of performance. The size of the pressurized water-cooled nuclear reactor developed for the Soviet Navy dictated the minimum diameter of the pressure hull. However, the penalties incurred in adopting double-hull construction meant that a single reactor with an output comparable to that of the US Navy's 55W would drive the 'November' at a maximum speed of around 20 knots – little more than contemporary Soviet submarines with conventional propulsion machinery. The solution was to install two nuclear reactors, each with approximately the same power output as the single reactor in the US *Skipjack*. The adoption of a two-reactor design, however, necessitated a further increase in the length (and therefore the weight and resistance) of the hull.

As with the early American boats, the reactors were located roughly in the centre of the boat, where unexpected changes in weight during reactor development would have the minimum effect on trim. This placed the conning tower unusually close to the bow, forward of the reactor compartments.

Even with double the reactor output of the US Navy's *Skipjack*, the maximum speed of the 'November' was probably under 30 knots. Despite Soviet attempts to reduce resistance underwater by adopting a modern

Right: *A submarine of the 'November' class in distress off the West Coast of Ireland in 1970. She later sank. The early Soviet nuclear boats quickly established a reputation for poor reliability, with frequent breakdowns, fires and radioactive leakage. (US Navy)*

Below: *Early units of the 'November' class had a conning tower with a vertical face, into which nine passive sonar panels were set. Later units had a more streamlined conning tower with a sloping forward edge which was subsequently adopted for the 'second generation' SSNs of the 'Victor' class.*

'tear-drop' hull-form topped by a small, streamlined fin, there were features which militated against high underwater speed.

The large number of free-flood apertures in the outer casing – an unavoidable feature of double-hull construction – increased resistance when the boat was running at speed underwater. Any attempt to reduce the number or size of these free-flood holes would have led to an unacceptable increase in the time it would take to dive.

The stern section of the 'November' was entirely conventional. The Soviet designers were unwilling to

'November'
1. attack periscope
2. search periscope
3. HF communications mast
4. Snoop Tray surface surveillance radar
5. Stop Light ESM
6. VHF/UHF communications mast
7. Quad Loop DF

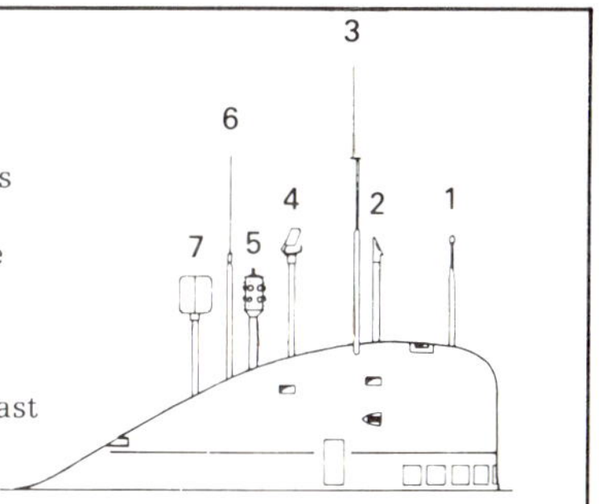

'NOVEMBER' CLASS

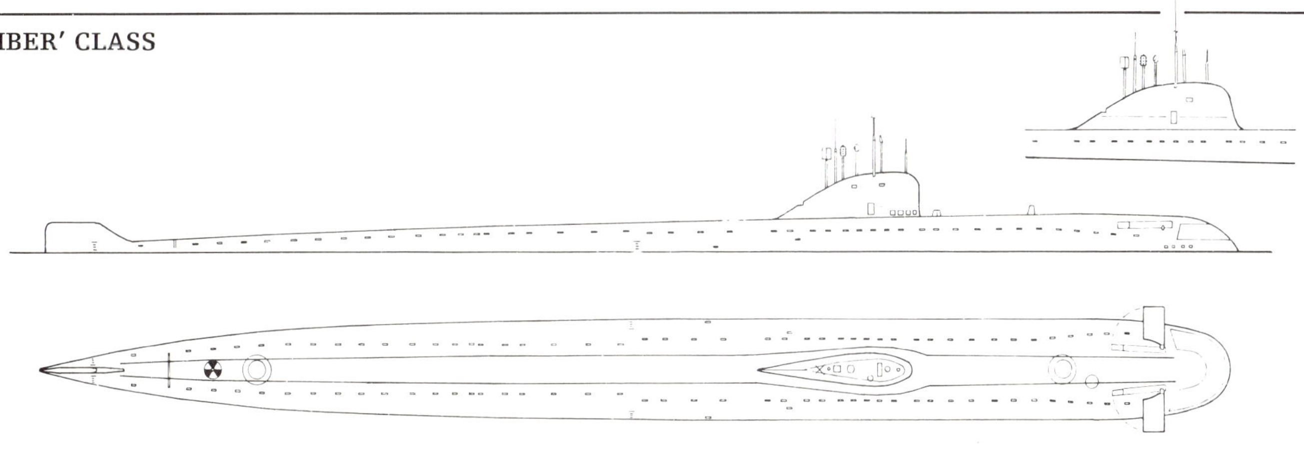

Left: *One consequence of the adoption of a two-reactor nuclear propulsion plant is that the conning tower is placed well forward as compared with contemporary diesel-powered boats. Note the bridge windows, which are set into a hinged panel which fits flush with the conning tower casing. (Novosti, 1968)*

go to a single shaft, possibly because of concerns about reliability given such a high power loading on a single propeller. Instead they persisted with the traditional two-shaft arrangement, which again brought with it serious weight penalties and further reduced performance. The cruciform after control surfaces incorporated in the US Navy's *Skipjack* could only function effectively with a single shaft, so the Soviet boats were fitted instead with a conventional rudder and separate horizontal diving planes. The short 'fin' which protrudes above the stern casing appears to be simply a form of vertical stabilizer, and has no moving parts.

Conservatism or safety concerns?

A number of reasons have been advanced for the retention of double-hull construction in Soviet submarines with nuclear propulsion.

Technological inertia was undoubtedly an important factor in Soviet persistence with the double hull for nuclear boats of the 'November' generation. Post-war reconstruction of the Soviet shipyards had barely been completed when the hull of the first 'November' was laid down. It was hardly the moment for a radical departure from established construction techniques. Externally framed pressure hulls were easier and less costly to build than internally framed single hulls. They were also inherently stronger and required less advanced steels. Moreover, there was more usable volume in an externally framed pressure hull than in a single hull of similar diameter. It was therefore theoretically possible to accommodate propulsion systems of comparable size and power in a smaller-diameter pressure hull if double-hull construction were employed.

The apparent haste with which the designs for the 'November' and its nuclear-powered contemporaries of the 'Hotel' and 'Echo' classes were drawn up makes it unlikely that the Soviets would have had sufficient time to evaluate fully the relative merits of single- and double-hull construction, and the desire to get the boats into service at the earliest possible date effectively precluded any departure from traditional shipyard building practices. The retention of the double hull for these early boats is therefore unsurprising.

There is evidence to suggest, however, that there were also positive reasons for persisting with double-hull construction, reasons that were to become clearer during subsequent discussion in Soviet professional and technical journals. During the mid-1960s, by which time the second generation of Soviet nuclear boats was on the slipways, a series of officially endorsed articles and books appeared, all of which argued against the adoption of single-hull design.[2]

It appears that the Soviet designers were particularly concerned about the reduction in reserve buoyancy (from 25–35 per cent of surfaced displacement in a double-hull design to 9–13 per cent in a single-hull design) associated with single-hull construction. This reduction would limit the amount of flooding a submarine could overcome in the event of action damage. Moreover, the distribution of the ballast tanks along the length of the boat, which was standard practice in a double-hull design, was considered superior. Single-hull boats had their main ballast tanks at their fore and after extremities, where their integrity could easily be breached by a collision.

A further concern was the reduction in compartmentation in Western single-hull designs, which generally have only three watertight compartments as compared to six in the Soviet boats.

An additional protective feature of double-hull design is the stand-off between the inner and outer hulls. In theory much of the force of a conventional HE torpedo warhead would be expended on the outer hull casing of the submarine, making it less likely that the pressure hull would be breached. This argument was lent some weight by the widespread use of small-diameter (324mm) air- and ship-launched anti-submarine homing torpedoes in Western navies. The 45kg warhead of these torpedoes was barely capable of disabling a submarine with double-hull construction, being only one-third the weight of the warhead carried by the 533mm Mk 37 ASW torpedo then in service with US Navy submarines.

The space between the two hulls was also useful for the installation of weapons and sensors. Soviet nuclear-powered submarines from the 'Echo' onwards utilized this space to accommodate missile launch tubes, and later boats would stow a variety of equipment, from communications buoys to towed arrays, between the two hulls. In single-hull Western boats these items of equipment frequently had to be stowed within the main ballast tanks.

The Soviets appear to have been fully aware of the penalties they would have to pay if they persisted with double-hull construction. The writers of the articles and books that appeared in the 1960s, however, were of the opinion that many of the disadvantages inherent in double-hull construction could be mitigated by future technological developments. It was thought that speed loss resulting from the adoption of larger overall dimensions could be compensated by the development of high-density reactors, which would give increased power for a given size. Diving depth could be increased and overall structural weight reduced by the adoption of high-tensile steels, and the use of lightweight alloys in both the pressure hull and other structures.

It was felt that quieting could be improved by superior outer hull design and by the resilient mounting of equipment, although given the projected missions of Soviet submarines of the time quieting clearly had a much lower priority than was the case for contemporary Western designs, which were intended primarily for anti-submarine warfare.

'NOVEMBER' CLASS

Soviet designation: Podvodnaya Lodka Atomnaya (PLA) (Submarine, Nuclear-Powered)

Construction:
Built at Shipyard No. 402, Severodvinsk. Fifteen units completed 1958–64; one in 1958, two in 1959, one in 1960, three in 1961, three in 1962, three in 1963, two in 1964.

Displacement:
Surfaced: 4,500 tons
Submerged: 5,300 tons

Dimensions:
Length: 110m (361ft)
Beam: 9m (30ft)
Draught: 7.7m (25ft)

Propulsion:
Two pressurized water-cooled reactors driving geared steam turbines; two shafts with 4- or 6-bladed propellers; 30,000shp for 28 knots max.

Complement 80

Weapons:
Eight bow tubes of 533mm (21in) diameter for anti-ship torpedoes; four stern tubes of 400mm (16in) diameter for anti-submarine torpedoes; maximum capacity twenty-four 533mm torpedoes, eight 400mm torpedoes.

Sensors:
MF active/passive sonar; fin-mounted passive FC arrays; search and attack periscopes; Snoop Tray surface surveillance radar; Stop Light ESM; Quad Loop DF.

Communications:
HF and VHF radio.

Nuclear torpedoes

The fully rounded bow section of the 'November' made it possible to accommodate no fewer than eight torpedo tubes of standard 533mm diameter in addition to the main passive sonar array, and the handling room abaft the tubes was of sufficient size to accommodate an estimated sixteen reloads – two per tube.

A number of commentators have advanced the theory that the 'November' was designed from the outset as a platform for the nuclear torpedo which began development in the mid-1950s, and is thought to have entered service in 1959. The main body of the torpedo appears to have been based on the standard Type 53-VA steam torpedo, although at least one commentator has stated that it was based on the later HTP-fuelled variant discussed in an earlier chapter.[3]

A torpedo with a nuclear warhead has the advantage of a much larger lethal radius than one armed with conventional high-explosives, and is therefore particularly attractive for use against large formations. However, it needs to be launched from a considerable distance if the resulting explosion is not to disable the

Left: *The* Leninsky Komsomol *was the first Soviet submarine to reach the geographic North Pole, where this photograph was taken in 1962. Despite the notorious defects of the early nuclear boats, their exploits were much lauded in the Soviet Press, and command of a 'November' could do much to enhance a serving officer's career prospects. Admiral Chernavin himself commanded 'November'-class submarines of the Northern Fleet during this period. (Novosti)*

submarine itself, and this creates guidance problems. The US Navy's Mk 45 ASTOR, which entered service at about the same time as the Soviet model, employed wire guidance, but was designed for use against submarines and therefore needed to be extremely accurate to compensate the imprecisions in targeting inherent in underwater combat.

The Soviet weapon, which was designed for surface targets, was almost certainly a straight-running weapon, and was probably detonated after a pre-determined run-out employing a propeller turn count. The original purpose of the weapon is thought by some commentators to have been strategic, the nuclear torpedo being intended for use against Western port installations. However, the value of such a weapon against large concentrations of surface ships quickly became apparent, and there is evidence to suggest that deployment of the nuclear torpedo was extended to the conventionally powered 'Foxtrot', and even to the medium submarines of the 'Whiskey' and 'Romeo' classes.

The tactical advantage held by the 'November' over its diesel-powered counterparts lay in the sustained high underwater speeds possible with nuclear propulsion. Whereas the diesel boats would need to form a patrol line across the expected path of advance of an approaching task force, the 'November', operating as a 'lone wolf', could be vectored towards the task force from almost any angle, using its high speed to close the enemy from the flank and take up a favourable firing position.

Unfortunately, the 'November' proved to be an extremely noisy boat, particularly when operating at high speed, and as Western sonar capabilities improved during the 1960s it became a relatively easy matter to detect its approach. There were a number of features in the design which contributed to noisy underwater operation. The free-flood holes which lined the outer casing, and were a necessary feature of double-hull design, contributed to flow-noise. More significantly, the twin propellers created massive water disturbance when the 'November' operated at higher speeds. Nor was there any conscious attempt to reduce machinery noise. The steam turbines and their associated gearing were mounted directly onto the hull, as was all auxiliary machinery including the pumps responsible for circulating the coolant water through the reactor chamber.

It is only fair to point out that some of these features were also common to the early nuclear submarines built for the US Navy, but quiet operation was already a major consideration in the design of new submarine types for the US Navy during the late 1950s, whereas the Soviets showed no interest in quieting measures at this time. Even the streamlining of the hull and fin which was such a distinctive feature of the 'November' class was related to underwater performance; the Soviets were far more concerned about water resistance, which detracted from high speed, than they were about flow noise.

Anti-submarine torpedoes

All the major reference sources credit the 'November' with four short stern tubes of 400mm diameter for anti-submarine torpedoes. The adoption of small-diameter tubes may have been influenced by the hull-form of the 'November', which featured a fine, tapered stern section. Similar considerations were responsible for the short 'anti-escort' torpedo tubes in some Western post-war diesel submarines (notably the US *Tang* and the British *Porpoise* classes).

The different function of the short torpedoes carried by the 'November' – anti-submarine, not anti-escort – reflects important differences in tactical thinking. The Western diesel boats were intended to fire their conventional torpedoes from close in, to be sure of hitting their target. The major threat to their own safety was posed by the destroyers or frigates escorting the target. The 'November', on the other hand, would not attempt to break through the anti-submarine screen, but would launch its nuclear torpedoes from a safe distance to avoid detection by the escorting destroyers. The major threat would be from hunter-killer submarines operating in support (direct or indirect) of Western surface shipping. If an enemy submarine were detected, the 'November' would turn away, launch its torpedoes from the stern tubes, and make off at high speed.

Tactics such as these favoured a free-running weapon with active acoustic homing, wire guidance being inherently unsuited to high-speed manoeuvres. The Soviet torpedo fired from the stern tubes of the 'November' is reportedly just such a weapon, and has since been deployed on surface vessels such as the Guard Ships (SKR) of the 'Petya' and 'Mirka' classes. There is some disagreement regarding the date the torpedo entered service, although it would seem only logical to conclude either that it entered service with the 'November' class, or that it was available within two years of the completion of the first unit in 1958. The torpedo, designated SET-40 in some sources,[4] has a length of about 4.5m and electrical (battery) propulsion, and can be deployed in the swim-out mode from standard 533mm tubes in other classes, using a torpedo tube liner.

Sonars

The 'November' differs from its diesel-powered contemporaries in having no sonar dome atop the bow. There is, however, a prominent acoustic-transparent 'window' around the bow itself. The near-hemispherical configuration of the bow suggests that it may house a passive array of significantly greater diameter than the 2–2.5m array of the 'Foxtrot'. This appears unlikely, given the haste with which the 'November' was designed and built, but the possibility cannot be discounted. The array may have served as a prototype for the large low-frequency array incorporated into the later 'Victor'-class submarines. Given the absence of a separate above-bow dome, it must be assumed that the active ranging sonar is located within the bow casing, possibly transmitting through the circular passive array.

Early boats of the class had nine of the one-metre square panels also noted on the 'Foxtrot' around the forward edge of the fin, which was vertical. Later boats had a modified fin with an angled forward edge and the square panels removed. This suggests that the system did not prove successful – it was not repeated on later classes – and that the vertical face of the original fin (presumably essential to the effective operation of the system) may have created water turbulence. Similar problems appear to have been experienced with the vertical face of the fin on the later 'Charlie'. The modified fin served as the prototype for the one adopted for the SSNs of the 'Victor' class.

The Arctic dimension

In July 1958 the USS *Nautilus* succeeded in reaching the geographic North Pole, and subsequently completed a 4-day, 1,800 nautical miles underwater voyage across the Arctic Ocean. This exploit was to be repeated later in the same year by the USS *Skate* (SSN-578) which, in addition, surfaced through the thin ice which makes up a large proportion of the Arctic ice pack during the summer months.

Exploratory voyages through Arctic waters by nuclear-powered submarines equipped with ever more sophisticated navigational devices and under-ice sensors were to become a regular feature of US Navy operations over the next few years.

The Soviet Navy was determined not to allow the Americans free play in an area so intimately associated with its own strategic interests, and in 1962 a submarine of the 'November' class, the *Leninsky Komsomol*, surfaced for the first time at the North Pole. Similar operations were subsequently undertaken on a regular basis by Northern Fleet 'Novembers'. These operations had the dual function of establishing a Soviet 'presence' in the Arctic, and of developing the necessary equipment and navigational expertise to allow the Soviet Navy to exploit these desolate waters for its own strategic purposes. Only with the creation of the ballistic missile submarine 'bastions' in the 1970s would the full benefits of these early, hesitant steps be realized.

Missile submarines

The other major new development under Krushchev was the submarine armed with missiles for land attack.

Access to German rocket technology and to the trials and test installations at Peenemünde provided the impetus for programmes to develop operational missiles for the Soviet armed forces. The Ground Forces were to have been the main beneficiary of these early programmes, and the Scud-A tactical missile attained operational status as early as 1957, but the Soviets were quick to see the strategic potential of missiles with a simple ballistic trajectory.

Three major problems needed to be overcome if this potential was to be realized. The early land-based rockets had insufficient range to threaten the countries of Western Europe, let alone the United States. They were also very inaccurate, as there was no possibility of mid-course guidance, and the probable miss distance (or C.E.P.) would inevitably increase with longer range. Moreover, the early rockets were incapable of lifting a nuclear warhead, which could otherwise have compensated any inaccuracies in the inertial guidance system. The necessary increase in range and payload could be resolved by the development of larger, more powerful rockets, but this solution appeared to preclude the employment of surface vessels or submarines as a launch mode.

Missile development in the United States followed a different pattern. During the early post-war period US submariners were particularly anxious to acquire a missile launch capability. The Soviet Navy presented few inviting surface targets, and the Soviet merchant marine was small and strategically insignificant. The missile land attack mission therefore seemed particularly attractive as a means of enhancing both the utility and the prestige of the submarine arm.

The US Navy could see no short-term solution to the technical problems presented by the ballistic missile, and therefore opted for the development of cruise missiles derived from the German V-1, which was essentially a pilotless aeroplane powered by an air-breathing turbojet. A cruise missile could carry a heavier payload of explosives than a lightweight ballistic rocket, but its major advantage was that it was susceptible to guidance. Its greater potential accuracy made it appear more promising as a long-range weapon, particularly if a conventional payload were carried. Various developmental projects were therefore pursued, resulting in an operational missile (Regulus I) as early as 1954.

The Soviets, on the other hand, opted from the outset for the development of ballistic rockets based on the German V-2. This choice was almost certainly influenced by a Soviet perference for land-basing, which made possible the future development of large long-range missiles employing vertical launch. Moreover, the guidance technology developed by the US Navy for its early cruise missiles owed little or nothing to German scientists; the German V-1 was simply pointed in the direction of its target, and fell out of the sky when it ran out of fuel. Rather, the guidance systems of the early US missiles were based on superior US electronics technology to which the Soviet Union had limited access.

However, the development of an operational 'intercontinental' ballistic missile capable of reaching US cities from the Soviet land mass was clearly some years away. The Soviets therefore needed a sea-based system which could be deployed sufficiently close to the American east coast to compensate the range deficiencies of their early rockets. Such a system was deemed essential if the US land mass were to be subjected to a threat similar to that posed to the Soviet Union by forward-deployed nuclear weapons based in Europe.

The technical basis for the prototype Soviet sea-launched ballistic missile (SLBM) system was provided by the German Lafferentz project, which was begun in 1944 but subsequently had to be abandoned because of the military situation in the Baltic. Under this project, V-2 rockets were to have been towed behind Type XXI submarines in watertight containers. When the launch

Left: *A 'Zulu V' returns to its Northern Fleet base following a succesful missile firing. Note the enlarged conning tower as compared with the 'Zulu IV'. (Novosti 1966)*

Right: *A close-up of the conning tower of a 'Zulu V' taken in 1973. The hemispherical caps for the twin launch tubes are clearly visible at the after end. The protruding 'chimney' is for the diesel exhaust. (US Navy)*

point was reached, the containers were brought to the vertical and ballasted by means of trim tanks controlled from the submarine. A party from the submarine would then prepare the rocket for launch; the nose cap of the rocket would be opened up, and the liquid fuel would be transferred from tanks in the container. The party would then withdraw to the submarine and the rocket motors would be fired under remote control.

Clearly this launch mode posed a number of serious problems, not least of which was the 30-minute preparation period during which the submarine would have to lie motionless on the surface. The relatively short range of the V-2 missile meant that the submarine would have to take up a firing position close to the enemy coastline if it were to fire against inland targets. Moreover, the effect of wave motion on the capsule, together with the difficulty in determining the precise position of the launch submarine, would ensure that the missile suffered from even greater inaccuracies than its land-based counterpart.

Nevertheless, there is evidence to suggest that the Soviets conducted similar experiments with containerized T-1 missiles (the Soviet development of the German V-2) in the Baltic during the early 1950s,[5] before abandoning this launch mode in favour of missiles fired from vertical launch tubes installed in the submarine itself.

The subsequent history of Soviet SLBM development is unclear, and attempts to trace the progression from the early V-2 derivatives to the SS-N-4, the first operational Soviet submarine-launched missile, have only served to add to the confusion. Some commentators have stated that rockets of the Golem and the Komet series played a part in the developmental process, and the Scud-A has also been cited as a possible development vehicle. Much of this is supposition, however, and there is little hard evidence. Missile tubes installed in submarines are necessarily of fixed diameter, and although it is possible to accommodate a smaller missile in the tube by the insertion of a liner, the Soviet missiles would have also required handling apparatus of different design given that their early SLBMs appear to have been launched clear of the tube from a cradle.

The first Soviet ballistic missile submarines were conversions of existing types. Six units of the 'Zulu' class were taken in hand and the fin was rebuilt to accommodate two launch tubes. The first of these conversions, which subsequently received the designation 'Zulu V', was completed in 1956, and the last was completed in 1958. According to one source,[6] these conversions were preceded by a trials conversion with a single missile tube. All conversions were carried out at the Severodvinsk Shipyard.

The basic hull-form of the 'Zulu V' remained unmodified, but the fin itself was five metres longer and more than two metres higher than that of the standard attack boat. It was also broader, in order to accommodate the two launch tubes, each of about 2.25m diameter, at its after end. No attempt was made to streamline the top of the fin, which was open with the missile tubes covered by hemispherical hinged caps (*see* drawing). The sensor masts were regrouped forward and aft of the missile tubes.

The effect of the two launch tubes on internal arrangements must have been considerable, given that no attempt was made to lengthen the hull. Most sources agree that the four stern torpedo tubes were retained, but it seems likely that the after torpedo

'ZULU V' CLASS

Soviet designation: Podvodnaya Lodka Raketnaya Ballisticheskaya (PLRB) (Ballistic Missile Submarine, Diesel-Powered)

Conversions:
Six units converted at Shipyard No. 402, Severodvinsk 1956–8.

Displacement:
Surfaced: 2,100 tons
Submerged: 2,600 tons

Dimensions:
Length: 90m (295ft)
Beam: 7.5m (25ft)
Draught: 6m (20ft)

Propulsion:
Diesel/electric drive on three shafts; three Type 37D diesels, 6,000bhp for 18 knots surfaced; three electric motors, 5,300hp for 15 knots submerged.

Complement 85

Weapons:
Two SS-N-4 Sark ballistic missiles in vertical launch tubes; six bow tubes, four stern tubes, all of 533mm (21in) diameter, for anti-ship torpedoes; maximum capacity eighteen torpedoes.

Sensors:
HF passive sonar; search and attack periscopes; Snoop Plate surface surveillance radar; Stop Light ESM; Quad Loop DF.

Communications:
HF and VHF radio.

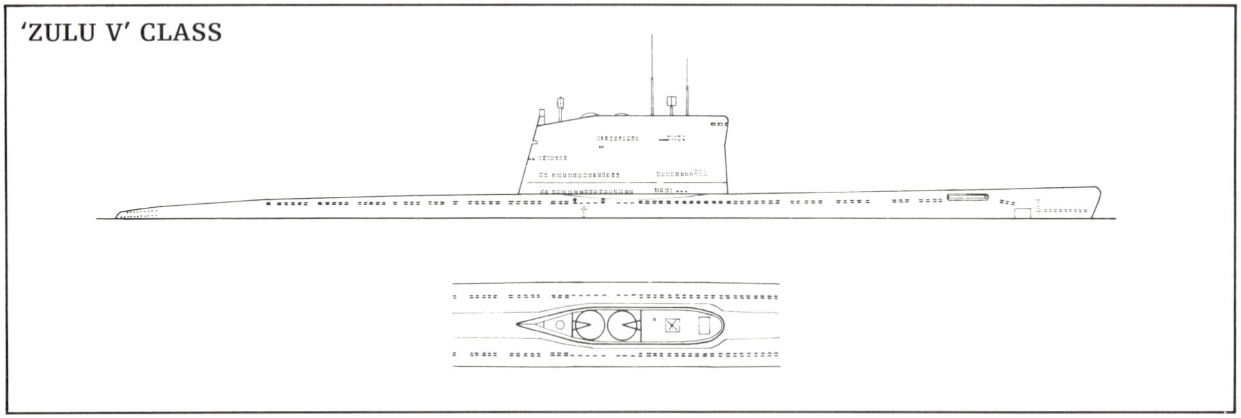

Left: *Note the circular configuration of the missile hatches in the plan view.*

handling room would have been utilized to extend the accommodation available for the crew, which included an additional contingent estimated at fifteen men for the maintenance and launching of the missiles.

The 'Zulu V' was clearly only a 'stopgap' conversion of limited military value. No attempt was made to modernize these submarines by equipping them with modern sonars, and they were discarded early in the 1970s. However, they provided the necessary experience for the development of purpose-built ballistic missile submarines which were to enter service during the late 1950s.

Severodvinsk, Komsomolsk and the 'strategic' factor

The first two units of the 'Golf' class were completed in 1958. Series production followed, with between six and eight boats entering service over each of the next three years. Construction then ended fairly abruptly, with a single unit completing in 1962.

Of the total of 23 'Golfs' completed, sixteen were built at Severodvinsk. The other seven units were built at the Leninskaya Komsomola Shipyard at Komsomolsk in the Far East. The choice of these two yards for such an important programme was significant. They were

Below: *A submarine of the 'Golf I' class. Note the massive conning tower, built to house three launch tubes for the SS-N-4 missile. (US Navy, 1962)*

Right: *A 'Golf I' underway in the Pacific in 1961. The second of the three SS-N-4 launch tubes has its cover hinged back to port. Note the tapered bow, yet to receive the Feniks passive sonar array.*

the only major Soviet shipyards located within the two fleet areas having direct access to the Atlantic and Pacific Oceans respectively. It was therefore essential that they be developed to enable them to build and to maintain submarines designed for a strategic mission. Construction facilities accordingly underwent considerable expansion during the early 1950s.

Severodvinsk was also the first Soviet shipyard to undertake the construction of submarines powered by nuclear reactors. The fifteen nuclear attack boats of the 'November' class were completed there from 1958 to 1964, at a building rate of two or three per year. This was a formidable achievement for a shipyard which had previously built only a handful of 'Zulu'-class boats, and illustrates the drastic reorientation of submarine building that took place in the Soviet Union during the 1950s. It also serves to underline the essentially *strategic* nature of the mission for which the 'November' was initially intended.

Only one year after the completion of the first units of the 'Golf' and 'November' classes, the lead-vessel of a third major type of strategic submarine, also built at the Severodvinsk Shipyard, entered service. The 'Hotel' did not enter production at Komsomolsk, which instead was to be responsible for the design and construction of a third nuclear-powered type armed with cruise missiles, the 'Echo', which entered service from about 1960.

The 'Golf'

The 'Golf' was essentially an enlarged 'Foxtrot'. It had an identical three-shaft propulsion system, and the same arrangement of torpedo tubes. The major difference lay in the massive fin, which was designed to accommodate three vertical-launch missile tubes. The fin was broader than that of the reconstructed 'Zulu V', and had a length of fully 26 metres as compared with only 19 metres for the latter. This enabled the designers not only to accommodate an additional missile tube, but to provide sufficient space for a more satisfactory layout of sensor masts forward of the missile hatches. The latter, unlike the hemispherical caps fitted over the missile tubes of the 'Zulu V', were integrated into the fin casing and hinged to port. All external protrusions, including the exhaust uptake at the after end of the fin, were fully retractable in order to improve the hydrodynamic flow. In spite of these measures the large wetted surface area of the fin was responsible for a significant reduction in underwater speed from a maximum of 15.5–16 knots in the 'Zulu' and 'Foxtrot' to about 12 knots in the 'Golf'.

'GOLF I' CLASS

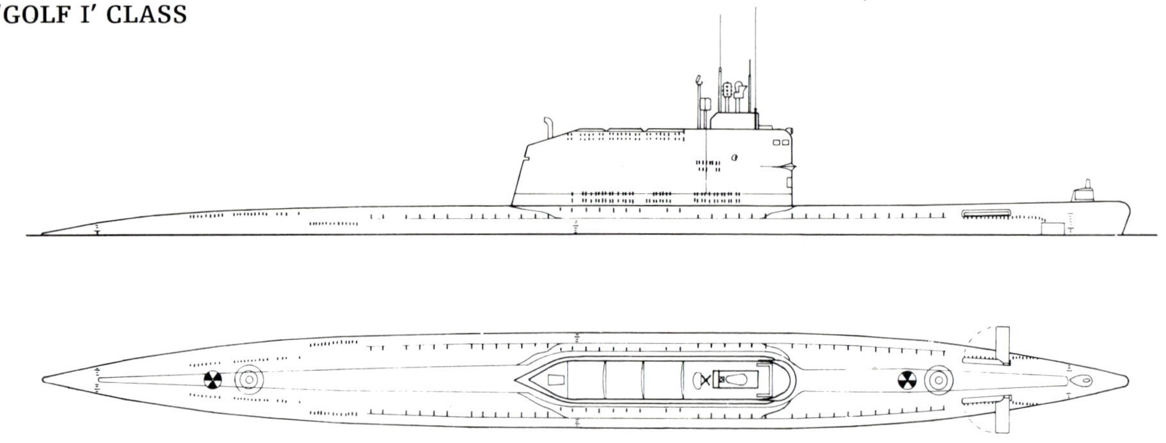

Left: *The drawings are of the 'Golf I' as originally completed.*

'Golf I'

1. HF communications whip
2. attack periscope
3. snorkel
4. Stop Light ESM
5. VHF/UHF communications mast
6. Quad Loop DF
7. Snoop Tray surface surveillance radar
8. diesel exhaust (retractable)

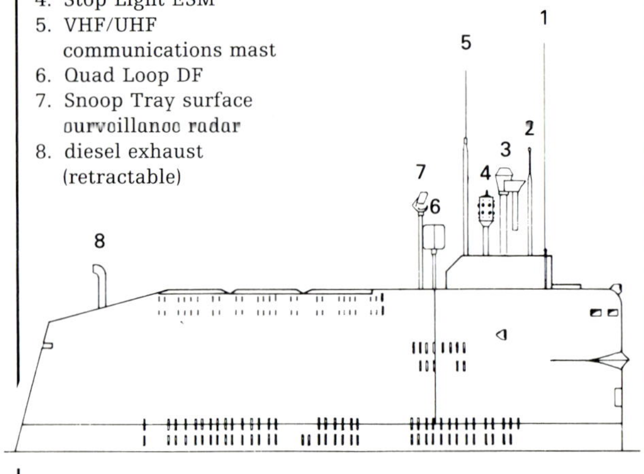

'GOLF' CLASS

Soviet designation: Podvodnaya Lodka Raketnaya Ballisticheskaya (PLRB) (Ballistic Missile Submarine, Diesel-Powered)

Construction:
Built at Shipyard No. 402, Severodvinsk; Komsomolsk-na-Amur.
23 units completed 1958–62; two in 1958, eight in 1959, five in 1960, seven in 1961, one in 1962.

Displacement:
Surfaced: 2,300 tons
Submerged: 2,700 tons

Dimensions:
Length: 100m (328ft)
Beam: 8.5m (28ft)
Draught: 6.6m (22ft)

Propulsion:
Diesel/electric drive on three shafts; three diesels, 6,000bhp for 17 knots surfaced; three electric motors, 5,300hp for 12 knots submerged.

Complement 80

Weapons:
Three SS-N-4 ballistic missiles in vertical launch tubes, six bow tubes, four stern tubes, all of 533mm (21in) diameter, for anti-ship torpedoes; maximum capacity eighteen torpedoes.

Sensors:
HF passive sonar; Herkules HF active sonar; search and attack periscopes; Snoop Plate or Snoop Tray surface surveillance radar; Stop Light ESM; Quad Loop DF.

Communications:
HF and VHF radio.

In order to accommodate the three missile tubes without reducing internal volume, hull length was increased by 10 metres, and in order to maintain stability there was a corresponding increase of one metre in beam. The complement was only marginally greater than for a 'Foxtrot', so there appears to be no reason why the number of torpedo tubes should have been reduced. However, the additional volume taken up by the purpose-designed missile control centre almost certainly meant that there were fewer reloads than were carried by contemporary attack boats.

A further important difference was that as initially completed the 'Golf' had a 'knife' bow not unlike that of the 'Zulu'. Only when the boats were converted to the 'Golf II' configuration did they receive the Feniks passive sonar array, characterized by the rounded bow section with its prominent unpainted 'window', which was fitted from the outset in the 'Foxtrot'. The early 'Golf I's had only the Herkules active sonar housed in an above-bow sonar dome and topped by an underwater telephone.

The first Soviet SLBM

There is still some confusion regarding the precise designation and configuration of the early Soviet submarine-launched ballistic missiles. The first SLBM to attain operational status was the SS-N-4, which entered service probably in 1959. Since the first 'Golfs' commissioned a year earlier, it seems likely that these boats conducted further trials with an interim missile derived from the army's Scud missile, as reported in some sources.

This missile was apparently designated R-11FM by the Soviets, and was considerably smaller than the SS-N-4. It may be the missile depicted in a rare photograph published on page 96 of Polmar's *Guide to the Soviet Navy* (4th edition), as the missile bears no resemblance to either the SS-N-4 or the later SS-N-5, even though the photograph is clearly of a submarine of the 'Golf' class, and not of the 'Zulu' class as stated in the caption.

The missile generally referred to as the SS-N-4 was a giant three-stage rocket with a conical nose-cone and seven prominent jet nozzles which made its first public appearance at the 1962 November parade marking the 45th anniversary of the Russian Revolution. With a length estimated at almost 15 metres and a diameter of about 1.8 metres, it could only just have fitted into the launch tubes of the 'Zulu V' and the 'Golf'. It is of interest to note that in 1956 a missile of even greater size was proposed for launch from US submarines. The Jupiter-5 had six rockets clustered together in its first stage and would have weighed about 70 tons. It was estimated that to carry four missiles a submarine of 8,500 tons would be required, the launch tubes being located in the fin as in the Soviet missile boats.

However, unlike the Jupiter missile the SS-N-4 had liquid-fuel propulsion, which had been firmly rejected by the US Navy as too hazardous for submarine operations. Soviet persistence with liquid fuel for its submarine-launched missiles was indicative both of a

Below: *A submarine of the 'Hotel I' class running on the surface. Pennant numbers were carried even by Soviet nuclear submarines until the late 1960s. (US Navy, 1967)*

Above: *Two SS-N-4 SLBMs are wheeled through Red Square for the 1964 May Day Parade.*

lower level of technology and of lower safety standards, and was to have disastrous consequences for at least one Soviet ballistic submarine.

A major weakness of the SS-N-4 missile was that it had to be fired from the surface, so that the submarine was exposed to detection and pre-emptive attack during launch preparations. The relatively short range of the missile – 300–350 nautical miles – meant that it would have to be launched close to the US coast, making detection even more likely. The SS-N-4 was therefore regarded by the Soviet Navy as only an interim solution to the problem of deploying nuclear missiles capable of targeting the US land mass. It constituted the first stage of a long-term SLBM development programme which would eventually deliver missiles capable of underwater launch at ranges in excess of 4,000 nautical miles and which could strike at targets throughout North America from patrol areas close to the Soviet Union.

That technology was still a long way off in 1959. Less than a year later, the US Navy would deploy the Polaris A-1 missile, which was capable of underwater launch, was powered by solid-fuel rockets, and had a range almost four times that of its Soviet counterpart. It was also less than two-thirds the length of the SS-N-4, enabling it to be accommodated comfortably within the hull of the submarine, in a missile section housing no fewer than sixteen missile tubes. It would be a further seven years before the Soviet Navy could deploy a submarine of similar configuration and capabilities.

The 'Hotel'

The US Navy required submarine-launched ballistic missiles of great range because the launch points available in the Barents and Norwegian Seas and the eastern Mediterranean were a considerable distance from the major Soviet cities, most of which were far inland.

Soviet SLBMs did not require such long range because many of the major centres of population in the United States were on or near the west or east coasts, but lengthy transits would be required in order to reach the favoured launch points. The ideal solution to

'HOTEL I' CLASS

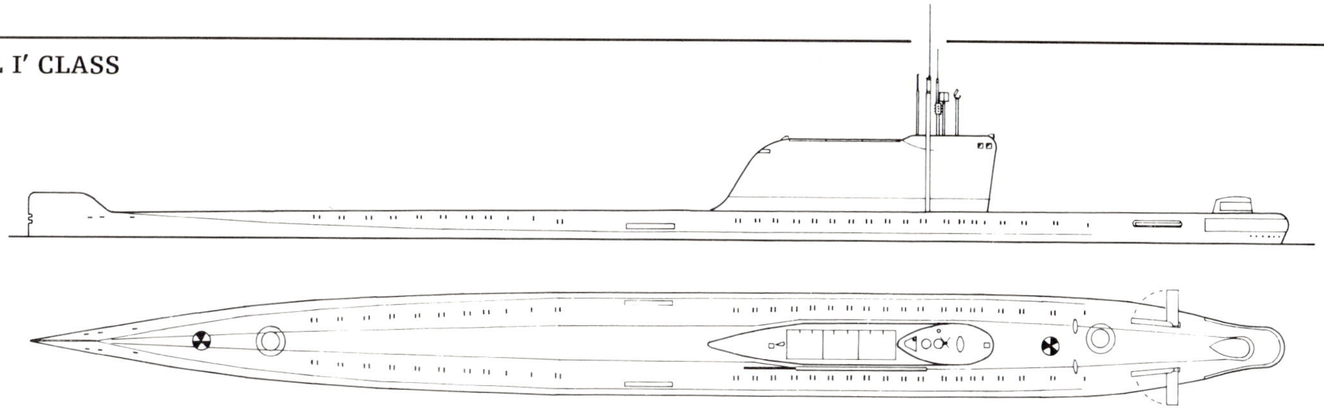

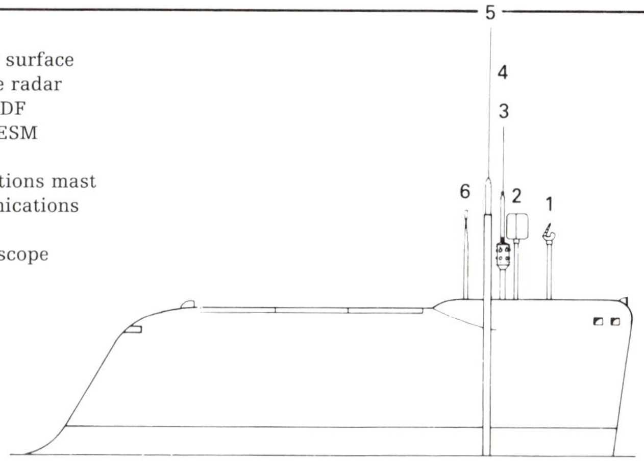

'Hotel I'
1. Snoop Tray surface surveillance radar
2. Quad Loop DF
3. Stop Light ESM
4. VHF/UHF communications mast
5. HF communications mast
6. search periscope

'HOTEL' CLASS

Soviet designation: Podvodnaya Lodka Atomnaya Raketnaya Ballisticheskaya (PLARB) (Ballistic Missile Submarine, Nuclear-Powered)

Construction:
Built at Shipyard No. 402, Severodvinsk. Eight units completed 1959–62: two in 1959, one in 1960, three in 1961, two in 1962.

Displacement:
Surfaced: 5,000 tons
Submerged: 6,000 tons

Dimensions:
Length: 115m (380ft)
Beam: 9m (30ft)
Draught: 7m (23ft)

Propulsion:
Two pressurized water-cooled reactors driving geared steam turbines; two shafts with 6-bladed propellers; 30,000shp for 24 knots max.

Complement 90

Weapons:
Three SS-N-4 ballistic missiles in vertical launch tubes; six bow tubes of 533mm (21in) diameter for anti-ship torpedoes; two stern tubes of 400mm (16in) for anti-submarine torpedoes; maximum capacity eighteen 533mm torpedoes, six 400mm torpedoes.

Sensors:
Feniks MF passive sonar; Herkules HF active sonar; search and attack periscopes; Snoop Tray surface surveillance radar; Stop Light ESM; Quad Loop DF.

Communications:
HF and VHF radio.

this problem was a submarine with nuclear propulsion, and a year after the first units of the diesel-powered 'Golf' class entered service, the first 'Hotel' was completed by the Severodvinsk Shipyard.

The 'Hotel' was essentially a 'Golf' powered by nuclear reactors. It was a larger boat, with a length of fully 115 metres and submerged displacement of 6,000 tons, but this increase in size was related exclusively to the need to accommodate two bulky pressurized water-cooled reactors in line, in order to generate the necessary power for a maximum underwater speed in excess of 20 knots.

The propulsion system of the 'Hotel' was identical with that of the attack submarines of the 'November' class. This placed the massive fin well forward as compared to the diesel-powered 'Golf'. The stern of the 'Hotel', which featured a prominent rudder-less tail-fin, also closely resembled that of the 'November'. The hull-form was otherwise conventional, the bow being tapered in the manner of the 'Golf' and the 'Foxtrot'. As a conventional hull-form was also adopted for the nuclear-powered cruise missile submarines of the 'Echo' class, the first of which was to enter service the following year, it seems logical to conclude that this feature was related to the requirement for both types to operate on the surface when conducting their lengthy preparations for missile launch.

The tear-drop hull-form adopted for the 'November' is well-suited to submerged operation at high speed, but has long been recognized as unsuited to running on the surface, when the submarine tends to 'porpoise'. A high degree of stability was presumably considered essential for the surface launch of both types of missile, hence the reversion to a conventional hull-form for the 'Hotel' and the 'Echo'.

The large fin of the 'Hotel' was more streamlined than that of the 'Golf'. The forward edge was angled in the manner of the 'Foxtrot', and the after edge was rounded and tapered. The 'Hotel' appears to have been

Left: *A submarine of the 'Hotel' class wallowing in the North Atlantic swell in 1972. Problems with the propulsion plant continued to plague Soviet nuclear boats of the 'HEN' generation. The unit depicted was eventually towed back to the Northern Fleet area for repairs. (US Navy)*

fitted from the outset with a prominent hinged mast for HF communications, which could be folded back into a recess set in the deck casing along the starboard side of the fin.

Unlike the 'Golf', the 'Hotel' was fitted from the outset with both the Feniks passive sonar and the Herkules active sonar. Reference sources also credit these boats with small-diameter stern torpedo tubes, which were also a feature of the 'November'. The 'Golf', which was based on the conventionally powered 'Foxtrot', is thought to have an identical torpedo tube arrangement to the latter, with six 533mm tubes forward and four 533mm aft. The 'Hotel' has the six 533mm bow tubes, but is thought to have only two stern tubes of 400mm diameter for anti-submarine homing torpedoes.

Cruise missiles

The Soviet Navy was slow to see the potential of the cruise missile. Such was the Soviet obsession with the futuristic V-2 that virtually all their early efforts were focused on the development of ballistic missiles for both land and naval use.

The US Navy, on the other hand, initially saw the air-breathing cruise missile as the natural successor to the manned bomber. The development of electronic guidance systems would enable the cruise missile to become an unmanned aircraft capable of delivering with a high degree of accuracy a larger payload of conventional explosives than a ballistic rocket. The advent of lightweight nuclear warheads (the 1,500kg Mk 5 bomb was tested in 1951) would make such a weapon even more attractive as a delivery system.

The cruise missile was particularly attractive for naval use because specially fitted submarines could be pre-positioned closer to shore to provide mid-course guidance. This facility would have been difficult to provide for land use, as the territory over which the missile would have to fly would be under enemy control. Hence the Soviet preference for the development of ballistic missiles with inertial guidance, which were better suited to deployment with the Ground Forces.

The US Navy first tested the ramjet-propelled Rigel missile, with a range of 500 nautical miles, in 1950 and the turbojet-propelled Regulus, initially with a range of only 200nm, in March of the following year. Rigel eventually had to be cancelled, because the requirement for a 125-metre launch ramp precluded firing the missile from submarines. Regulus, on the other hand, could be launched from a 'zero-length' ramp, and

Above: *USS* Halibut *was to have been the first of a series of nuclear-powered submarines designed to launch the Regulus nuclear-tipped cruise missile against the Soviet land mass. This development provided much of the impetus for the development of similar cruise missile submarines by the Soviet Navy. Note the different arrangements for handling the missiles, which were broken down for stowage in the large 'hump' on the foredeck. (US Navy, 1968)*

achieved its first shipboard launch in November 1952. During the development of the missile, range was increased to 500nm, enabling it to strike at targets 200nm inland with the launch submarine positioned 300nm off the coast – the minimum distance considered acceptable to ensure the submarine's safety.

The first US submarine to be fitted with Regulus was a converted warbuilt fleet boat, the *Tunny* (SSG-282), which conducted test firings from July 1953, the missile being considered operational by 1954. A further conversion, the *Barbero* (SSG-317), was subsequently approved (FY55), and two purpose-built diesel-electric submarines, *Grayback* (SSG-574) and *Growler* (SSG-577) were duly ordered under the FY53 and FY55 programmes respectively. The converted fleet boats carried two Regulus I missiles; their new-built successors would initially carry four Regulus I, but were designed to accommodate Regulus II, a large supersonic missile with a range of 1,200 nautical miles then under development.

Regulus II was eventually to be cancelled in favour of Polaris, but not before the first five of a projected total of fourteen cruise missile submarines with nuclear power had been ordered. Four were subsequently to be re-ordered as ballistic missile submarines but *Halibut* (SSGN-587), which was at a relatively advanced stage of construction when the Regulus II missile was cancelled in 1958, was duly completed and entered service in 1960.

The four US Navy SSGs were transferred to the Pacific in 1959, and with the completion of *Halibut* they operated as four 'units' (the two converted fleet boats being counted as one unit), maintaining regular patrols on station until the first Polaris submarine assigned to the Pacific entered service in 1964. The two fleet boats were discarded shortly afterwards, and the three purpose-built submarines were converted to other roles.

Soviet developments

The history of the development of cruise missiles in the Soviet Navy provides an interesting comparison with these events.

First, it is clear that serious development did not begin until about 1954, after the death of Stalin. Nor can it be doubted that the major impetus came from Nikita Krushchev, who is alleged to have stated that any naval vessel incapable of launching a missile was not worth having. Once the decision was taken to develop cruise missiles for naval use, the new programmes were pursued with the same ruthless determination and urgency that characterized the Soviet drive to develop nuclear propulsion for submarines. Within a two-year period from 1958 to 1960 no fewer than three major cruise missile systems entered service with the Soviet Navy.

The medium-range SS-N-1 Scrubber missile equipped destroyer conversions of the 'Kildin' and 'Krupny' classes, while the short-range SS-N-2 Styx missile was deployed on fast patrol craft. Both carried a conventional explosive payload and employed radar homing, and they were intended for use against surface ships.

The third missile, the SS-N-3 (NATO code-name: Shaddock), was a strategic weapon similar in conception to the US Regulus, and was designed to be fired by submarines. With a range estimated at 400 nautical miles and armed with a nuclear warhead, it had a similar capability to the early Soviet submarine-launched ballistic missiles.

The first known installation of the SS-N-3 was aboard a medium submarine of the 'Whiskey' class, which was fitted in 1956–7 with a single cylindrical container on the after deck casing, giving rise to the NATO designation 'Whiskey Single Cylinder'. Intensive trials were conducted with this boat over the following two years while it was considered how best to deploy the missile.

The launch mode adopted for the SS-N-3 was very different from that developed by the US Navy for Regulus. In the US Navy cruise missile submarines, the missiles were broken down and stowed in a large hangar, which was integrated as far as possible with the outer hull casing. When the submarine surfaced, the missile had to be assembled on its launch ramp prior to firing.

The cylindrical container that housed the SS-N-3 on the Soviet boats doubled as a launcher. The launch cylinder was simply elevated by hydraulic rams to about 15–20 degrees for firing. The short, stubby wings of the SS-N-3 were folded against the body of the missile when stowed to enable the missile to be accommodated within the cylindrical container, and deployed only after launch. Like Regulus, Shaddock was powered by a turbojet, and launch was assisted by two solid-fuel JATO boosters slung beneath the body of the missile.

This was clearly a neater solution to the problem of accommodating the missiles in a submarine, as it eliminated the need to assemble the missile prior to launch, which could be a protracted and difficult operation in heavy seas. It also provided a launcher that was weather-proof almost up to the moment of

Below: *A 'Whiskey Twin Cylinder' dressed overall for Soviet Navy Day at Sevastopol in 1967. Note the prominent bulwark built out over the starboard saddle tank to protect the launch cylinder for the SS-N-3 missile.*

Right: *A 'Whiskey Twin Cylinder' prepares for missile launch. (Przemyslaw Budzbon collection)*

'WHISKEY TWIN CYLINDER' CLASS

Soviet designation: Podvodnaya Lodka Raketnaya Krylataya (PLRK) (Cruise Missile Submarine, Diesel-Powered)

Conversion:
Five units converted 1958–60.

Displacement:
Surfaced: 1,100 tons
Submerged: 1,400 tons

Dimensions:
Length: 76m (249ft)
Beam: 6.3m (21ft)
Draught: 5m (17ft)

Propulsion:
Diesel/electric drive on two shafts; two Type 37D diesels, 4,000bhp for 17 knots surfaced; two electric motors, 2,500hp for 10 knots submerged.

Complement 60

Weapons:
Two SS-N-3C Shaddock land attack missiles in elevating deck launchers; four bow tubes, two stern tubes, all of 533mm (21in) diameter, for anti-ship torpedoes; maximum capacity ten torpedoes.

Sensors:
HF passive sonar; Tamir 5 HF active sonar; search and attack periscopes; Snoop Plate surface surveillance radar; Stop Light ESM; Quad Loop DF.

Communications:
HF and VHF radio.

'WHISKEY TWIN CYLINDER' CLASS

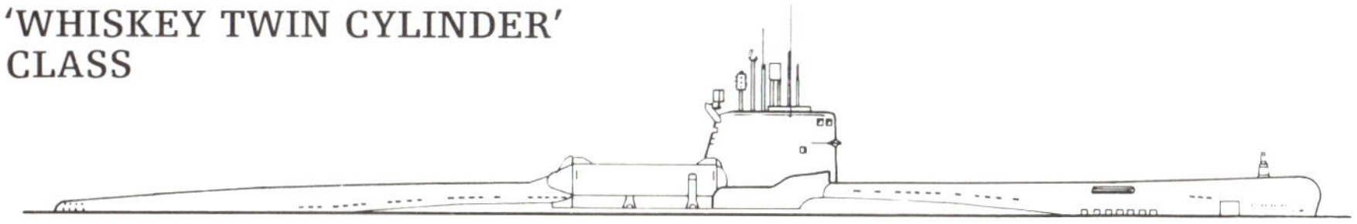

launching. This was an important consideration for the Soviets because of the unfavourable climate prevailing in the sea areas close to the USSR.

The major disadvantage with this mode of deployment was that a cylindrical deck launcher could not easily be integrated with the outer hull casing of the submarine. In the early 'Whiskey' conversions the launch tubes had to be added externally, with serious consequences for underwater performance.

The initial production version of the 'Whiskey' SSG was dubbed by NATO 'Whiskey Twin Cylinder'. Its twin launch tubes were located abreast immediately abaft the fin, facing aft. The seatings for the hydraulic elevating mechanism were located on the upper side of the protruding saddle tanks. As the missile launchers projected over the sides of the submarine and were quite close to the surface of the water, they were protected from the seas by prominent curved bulwarks angled out from the forward edge of the fin. These appear to have doubled as blast deflectors.

Few other significant modifications were made to the original 'Whiskey' design. The fin was 'tidied up', and a new high-frequency active sonar was added in a small above-bow dome. All the torpedo tubes were retained, but it is probable that the after torpedo room was used for additional accommodation, thereby reducing the number of reloads that could be carried.

The 'Whiskey Twin Cylinder' was a particularly

'ECHO I' CLASS (as SSGN)

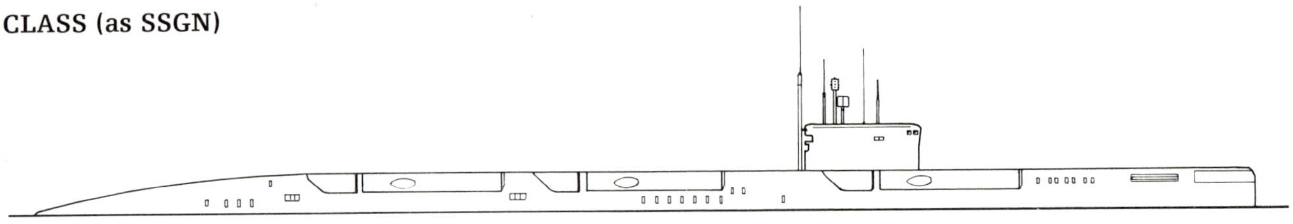

clumsy conversion, and only six boats were thus modified (from 1958 to 1960). Underwater speed suffered considerably from the various protrusions (the reduction of 3.5 knots from the original design cited in most sources is almost certainly an underestimate). On the other hand, flow noise increased, and one imagines that the handling of these boats when submerged must have left much to be desired. Nevertheless, the 'Whiskey Twin Cylinder' laid the foundations for a major programme of cruise missile submarines which was begun in the late 1950s.

The 'Echo'

The first submarine of the 'Echo' class was completed at Komsomolsk in 1960, which suggests that design work must have been complete by 1957, when the prototype 'Whiskey Single Cylinder' was still conducting trials with the SS-N-3 Shaddock missile.

The 'Echo' was a large nuclear-powered boat with overall dimensions similar to the 'November' and an identical propulsion system. The hull-form, however, like that of the 'Hotel', was entirely conventional. Integrated into the upper corners of the deck casing on each side of the boat were three pairs of elevating launchers for the SS-N-3 missile. Abaft each launcher were prominent indentations in the hull casing which served to deflect the blast of the missile exhaust outwards. The launchers themselves were elevated in pairs to an angle of about 15 degrees for firing.

There was a comparatively small, low conning tower of broadly rectangular configuration. As in the other early Soviet nuclear boats it was located well forward because of the length required for the machinery spaces, which featured two in-line reactors and geared turbines driving twin shafts. The curious rudderless 'fin' incorporated in the stern of the 'November' and 'Hotel' classes was not repeated on the 'Echo'.

There are no published photographs of the submarines of the 'Echo I' class in their original configuration, so that other details are more difficult to determine. They appear to have had a sonar fit similar to that of the 'Golf I', with a small above-bow dome for the Herkules active sonar and a cone for an underwater telephone. There is some disagreement in published reference sources concerning the torpedo armament, but there does not seem to be any compelling reason why this should have differed from that of their successors of the 'Echo II' class, which had six bow tubes of standard 533mm diameter and four short 400mm stern tubes for anti-submarine homing torpedoes.

Five submarines of the 'Echo I' class were completed at Komsomolsk from 1960 to 1962. It is interesting that the Soviets should have selected the Pacific shipyard to build this type rather than Severodvinsk in the Arctic. In part this decision may relate to a desire to give the two shipyards equal status as regards the construction of nuclear-powered submarines, as Severodvinsk appears to have been the lead yard for both the 'November' and the 'Hotel' classes. However, the

'ECHO I' CLASS

Soviet designation: Podvodnaya Lodka Atomnaya Raketnaya Krylataya (PLARK) (Cruise Missile Submarine, Nuclear-Powered)

Construction:
Built at Komsomolsk-na Amur.
Five units completed 1960–2: two in 1960, two in 1961, one in 1962.

Displacement:
Surfaced: 4,600 tons
Submerged: 5,400 tons

Dimensions:
Length: 110m (361ft)
Beam: 9m (30ft)
Draught: 7.5m (25ft)

Propulsion:
Two pressurized water-cooled reactors driving geared steam turbines; two shafts with 4-bladed propellers; 30,000shp for 25 knots max.

Complement 90

Weapons:
Six SS-N-3C Shaddock land attack missiles in paired elevating launchers; six bow tubes of 533mm (21in) diameter for anti-ship torpedoes; four stern tubes of 400mm (16in) diameter for anti-submarine torpedoes; maximum capacity twelve 533mm torpedoes, eight 400mm torpedoes.

Sensors:
Feniks MF passive sonar; search and attack periscopes; Snoop Tray surface surveillance radar; Stop Light ESM; Quad Loop DF.

Communications:
HF and VHF radio.

Left: *Conjectural drawing of the 'Echo I' as originally completed. There are no published photographs of this type in its original SSGN configuration.*

Right: *Despite the complete rebuilding of the centre section, the position of the sensor masts remains unchanged.*

'WHISKEY LONG BIN' CLASS

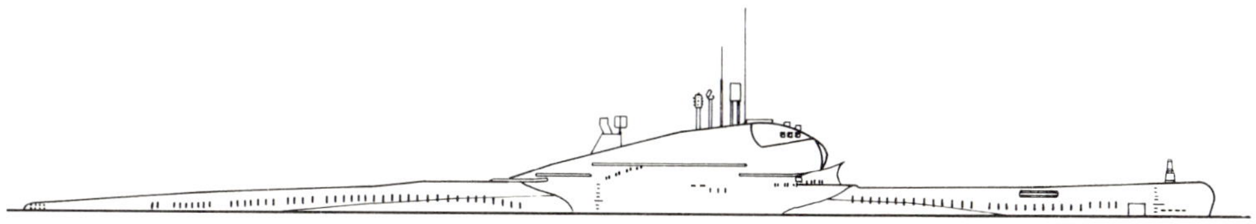

Soviets were undoubtedly aware of US intentions to deploy their own cruise missile submarines in the Pacific, and were anxious to create a similar force as a counterweight.

The 'Whiskey Long Bin'

The Soviet Navy under Gorshkov was now totally committed to nuclear propulsion for submarines, and to the dissemination of missiles throughout the fleet, but there was a limit to the number of nuclear-powered submarines that could be constructed in any one year in Soviet shipyards. Despite a massive redirection of resources, only the yards at Severodvinsk and Komsomolsk were as yet equipped to build nuclear boats. The available evidence suggests that by 1960 Severodvinsk could turn out a maximum of four nuclear-powered submarines per year – in itself a phenomenal achievement – while the capacity of the Komsomolsk Shipyard was only two.

The figure of six nuclear boats per year was inadequate to maintain a force structure capable of performing all Soviet submarine missions. Moreover, to confine submarine building to the Severodvinsk and Komsomolsk Shipyards would be unnecessarily restrictive, and would be a waste of the expertise built up at the Sudomekh Shipyard, which had been completely rebuilt post-war, and at the inland shipyard at Gorky. These shipyards would therefore continue to turn out conventional diesel boats to complement the nuclear construction at Severodvinsk and Komsomolsk. However, it appears that long-term plans to equip the Sudomekh and Gorky Shipyards to build nuclear-powered submarines were drawn up at about this time in an effort to boost production to at least ten nuclear submarines per year, a figure which was to be finally realized during the 1970s.

New classes of diesel submarine based on the torpedo-armed 'Whiskey' and 'Zulu' were already on the stocks by the late 1950s, and although this programme would subsequently be reduced in scale it could not be abandoned altogether. The shipyards concerned had spent a large part of the past decade tooling-up for this programme, and alternative designs were not available. Moreover, it was no easy matter to integrate guided missiles housed in bulky launch tubes, and requiring complex elevating mechanisms, into the design of relatively small submarines built around six or even ten torpedo tubes.

Redrawing the plans of the new types to accommodate such dramatic modifications would have caused considerable disruption to the production lines in the shipyards and would have delayed completion of the submarines. Furthermore, the Soviet Navy appears to have been generally satisfied with the capabilities of the 'Foxtrot', given that the nuclear-armed torpedo was about to enter service and that there was still a place for the long-range reconnaissance submarine within the new strategy.

The requirement for more submarine missile platforms was therefore met in typical Soviet fashion: a short-term 'contingency' programme involving the conversion of existing types, complemented by a longer-term programme for the design and construction of purpose-built missile boats.

'WHISKEY LONG BIN' CLASS

Soviet designation: Podvodnaya Lodka Raketnaya Krylataya (PLRK) (Cruise Missile Submarine, Diesel-Powered)

Conversion:
Six units converted at Leningrad 1960–2.

Displacement:
Surfaced: 1,200 tons
Submerged: 1,500 tons

Dimensions:
Length: 83m (274ft)
Beam: 6.3m (21ft)
Draught: 5m (17ft)

Propulsion:
Diesel/electric drive on two shafts; two Type 37D diesels, 4,000bhp for 13.5 knots surfaced; two electric motors, 2,500hp for 8 knots submerged.

Complement 65

Weapons:
Four SS-N-3C Shaddock land attack missiles in fixed angle launchers located in fin; four bow tubes of 533mm (21in) diameter for anti-ship torpedoes; maximum capacity eight torpedoes.

Sensors:
HF passive sonar; Tamir 5 HF active sonar; search and attack periscopes; Snoop Plate surface surveillance radar; Stop Light ESM; Quad Loop DF.

Communications:
HF and VHF radio.

Above: *A 'Whiskey Long Bin' photographed at Leningrad in the 1970s. The conning tower has been rebuilt around the four angled launch tubes for the SS-N-3 strategic cruise missile.*

The 'Whiskey Long Bin' was a much more fundamental reconstruction than the earlier 'Twin Cylinder' type. The 'Whiskey' hull was cut in half and had an additional section approximately eight metres long inserted. The fin was completely rebuilt to accommodate four fixed SS-N-3 launch tubes arranged as two pairs, the one superimposed above the other at an angle of about 10 degrees. Flaps were provided at the after end of the fin to allow the exhaust gases to escape.

In what was clearly an attempt on the part of the designers to overcome the hydrodynamic flow problems experienced with the 'Whiskey Twin Cylinder', the fin was low, broad and streamlined. Interestingly, the position of the mast-mounted sensors remained completely unchanged, except that the Quad Loop direction-finder, located atop a short 'chimney' for the snorkel exhaust, was farther aft, at the end of the new eight-metre hull section. This suggests that a new missile control room was housed in the new hull section, immediately abaft the original control room.

As with the 'Twin Cylinder', a small dome for an attack sonar and underwater telephone was added atop the bow, but only the four bow torpedo tubes were retained, the stern tubes being removed and the after torpedo room presumably being utilized for additional accommodation for the crew, which included a detachment estimated at fifteen men for the maintenance and firing of the missiles.

In spite of the efforts made to streamline the bulky fin of these boats, reference books generally credit them with a maximum submerged speed of only 8 knots – two knots fewer than the clumsy 'Twin Cylinder' modification. The sheer size and breadth of the fin, together with the weight of the missile launching equipment, must also have made them poor seaboats when running on the surface.

Strategic or tactical?

Given the limited endurance of the 'Whiskey' hull which formed the basis for these conversions, the intended mission of these submarines must remain open to question.

The SS-N-3C variant of the Shaddock missile carried by the 'Whiskey Long Bin' was incapable of guidance, and could therefore be used effectively only in the land attack role. Even then it was a less accurate weapon than the US Navy's Regulus missile, which could receive mid-course correction commands from other submarines posted closer to shore. The destructive radius of its nuclear warhead, however, was presumably considered sufficient by the Soviets to compensate any targeting inaccuracies. All that was required for a firing solution was accurate positioning of the submarine, plus the target coordinates.

The 'Whiskey' missile conversions all appear to have been carried out at Leningrad, but most units subsequently deployed to the Northern and Pacific Fleets, suggesting that their mission as initially envisaged was strategic, not tactical. However, the limited endurance of these submarines makes it more likely that they would have been deployed against targets in Europe and Japan, rather than the United States.

Some sources have stated that no fewer than 72 submarines of the 'Whiskey' class were to have undergone a 'Long Bin' missile conversion.[7] In the event only seven were modified, the programme having been drastically curtailed as a result of major policy changes which took place during 1959–60.

3. The Anti-Carrier Strategy

By 1960 the new submarine building programme instituted by Krushchev was in full swing. The last of the medium submarines of the 'Romeo' class had been laid down at Gorky, and a whole new range of large submarines had entered series production. The 'Golf'-class ballistic missile submarines were being built at a rate of 6–7 units per year at the Severodvinsk and Komsomolsk Shipyards, and their nuclear-powered counterparts of the 'Hotel' class at a rate of two per year at Severodvinsk. The first nuclear-powered submarines of the 'Echo' class, armed with the SS-N-3 Shaddock cruise missile, were coming off the production line at Komsomolsk, and a major SS-N-3 conversion programme was underway at Leningrad for the 'Whiskey'. The nuclear-powered 'November', armed with the new nuclear-tipped torpedo, was being completed at a rate of 2–3 units per year at Severodvinsk, and its conventional counterpart, the 'Foxtrot', was being built at a rate of 5–6 units per year, with construction shared between Severodvinsk and the Sudomekh Shipyard, Leningrad.

It was at this moment that yet another dramatic policy change occurred, with major consequences for the Soviet Navy's submarine building programme.

By the late 1950s the Soviet rocket programmes begun in the aftermath of the Second World War began to bear fruit. Large land-based ballistic missiles capable of reaching the United States from Soviet soil were now entering operational service. Major technical problems were experienced with the early missiles, so that only a handful were deployed from 1957 to 1961, but a second generation of Soviet ICBMs based on the SS-7 Saddler and the SS-8 Sasin would be ready for deployment by 1961. Successors to these missiles, the formidable SS-9 Scamp and SS-11 Sego, were already under development, and by 1966 production would reach two per day.

Land-based missiles, which were fired from hardened silos buried deep in the earth, had a number of important advantages over ballistic missiles launched from submarines. There were no constraints on the size of the missile, which could be made as large as it needed to be to cover ranges in the order of 4,000–5,000 miles. Moreover, a larger missile could carry a large warhead, capable of causing devastation over a broad area.

The other major factor was accuracy. Inter-continental ballistic missiles, in spite of the vast distances

Right: *A 'Golf II' conversion. The most obvious modifications are the Feniks passive sonar array in a new bulged bow, and the tall hinged mast for HF communications. There are also two square panels set into the forward edge of the conning tower similar to those fitted in the 'Foxtrot' and the 'November'. (MoD, 1979)*

they had to cover, were inherently more accurate than sea-launched missiles. Soviet first-generation ballistic missile submarines did not have effective inertial navigation systems, and surface launch of the missiles posed an insuperable stabilization problem in any sort of a sea. The inter-continental missiles, on the other hand, were fired from land-based silos occupying fixed coordinates. Targeting solutions were therefore relatively straightforward.

The year 1959–60 saw a major defence policy review during which all of these new technical developments were subjected to scrutiny. Krushchev seems to have been particularly concerned that a situation was developing whereby each of the Soviet armed services was attempting to carve out a nuclear role for itself in an attempt to sustain both status and funding. The Soviet Navy was embarked on its own submarine-launched ballistic missile programme, the Air Force on a major strategic bomber programme featuring air-launched nuclear missiles, and now the Ground Forces were set to deploy a new generation of inter-continental ballistic missiles.

These separate developments had been inevitable, given the largely experimental nature of most of the post-war military research programmes, and had been tolerated because of the need to pursue various alternative possibilities. Military programmes in the West were subject to the same constant reassessment during this period, and only stabilized during the 1960s.

The way forward, however, was becoming clear to the Soviet leadership. Given that strategic nuclear weapons were now the key factor in the overall military balance, it was vital that Soviet nuclear policy, the responsibility for which was currently distributed between the various armed services, should be unified. The essential element in a coherent nuclear strategy was central control over both development and deployment, and Krushchev felt that the only way that this could be achieved was by the creation of a separate branch of the armed forces, with responsibility for all long-range missiles. The Strategic Rocket Forces, which were formally set up in 1959, therefore took under their control all missiles with a range of more than 1,000 kilometres. Missiles of lesser range, which were designated 'operational-tactical' missiles by the Soviets, would continue to be assigned to the Ground, Air and Naval Forces.

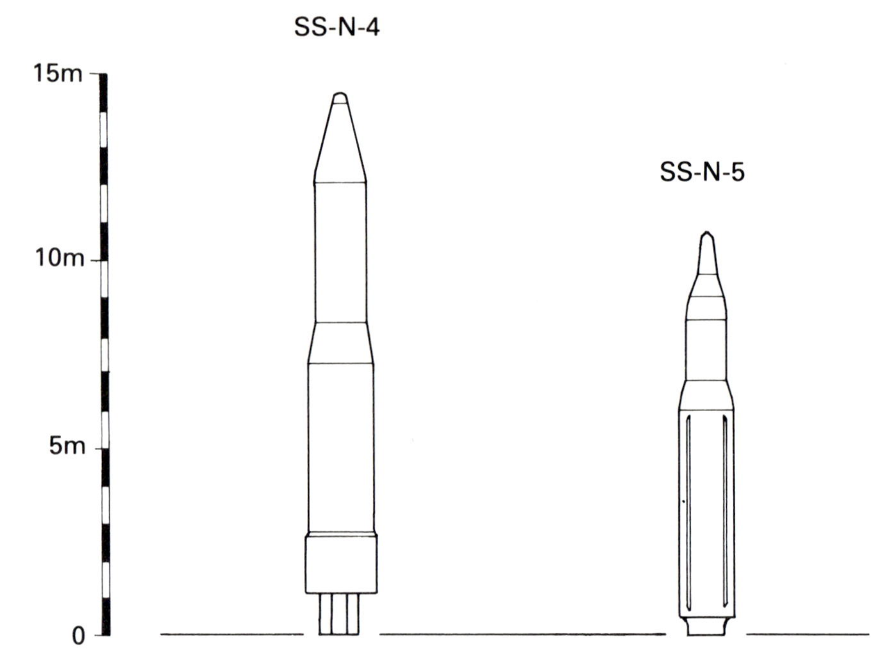

SS-N-4

Length:	14.5m
Diameter:	1.8m
Propulsion:	3-stage liquid-fuel rocket
Guidance:	inertial
Range:	350nm (650km)
Warhead:	1 RV × 800KT nuclear
In Service:	1958

SS-N-5 SARK

Length:	10.7m
Diameter:	1.5m
Propulsion:	2-stage liquid-fuel rocket
Guidance:	inertial
Range:	Mod. 1: 700nm (1,300km)
	Mod. 2: 900nm (1,650km)
Warhead:	1 RV × 800KT nuclear
In Service:	1963

Another new direction

The effects of these decisions on the Soviet Navy were every bit as momentous as the changes introduced by Kruschchev and implemented by Gorshkov following the death of Stalin. The submarine force, the principal fighting arm of the Navy, was in the process of being reconfigured for a strategic nuclear role, and now this central mission had been taken away.

No further units of the 'Golf' and 'Hotel' classes were laid down. The ballistic missile submarine programme was stopped in its tracks, and only those hulls still on the stocks were completed. The missile development programme was allowed to continue, and by 1963 some units of the 'Golf' and 'Hotel' classes were being retrofitted to fire the SS-N-5 missile, which had a range of some 700 nautical miles.

The SS-N-5 development programme was presumably already underway by the time the new policy changes were decided upon, and so was not subject to cancellation. The new missile was important to the Soviet Navy, not only because it had double the range of its predecessor, but because it could be fired from underwater. A cluster of eighteen small, electrically fired cold gas nozzles in a short section at the base of the rocket was used to eject the missile from its launch

Right: *As refitted, the 'Golf II' had the Feniks passive sonar array installed, a new hinged mast for HF radio communications was fitted abeam the forward end of the conning tower. By the late 1970s most units had been fitted with a deck housing for a VLF communications buoy. The drawings show the two major variants.*

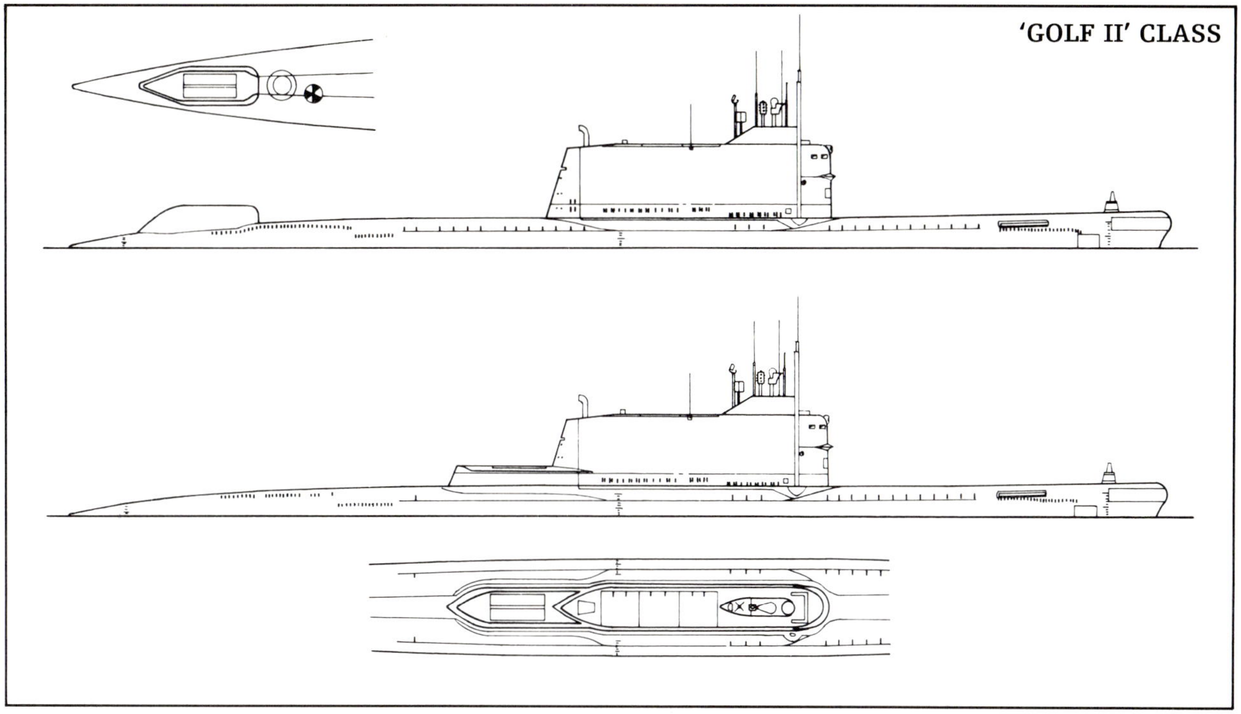

tube. This section was subsequently jettisoned by means of explosive bolts fired when the first stage ignited.

The SS-N-5 missile (NATO code-name: Sark) was of altogether more modern design than the SS-N-4, with only two stages and an external configuration not unlike that of the early Polaris missiles. It was also significantly smaller, with a length estimated by British sources at 10.7 metres and a diameter of about 1.5 metres.

The small size of the SS-N-5 is surprising. While it is a relatively simple matter to accommodate a smaller missile in an existing launch tube by inserting a 'sleeve', it is difficult to see why the Soviets should have sacrificed, for no apparent gain, the additional range which would have been possible with a larger missile employing the same technology.

It is therefore possible that the SS-N-5 was not designed primarily for retro-fitting in the 'Golf' and the 'Hotel', but was intended for a new class of submarine with launch tubes in the hull, modelled on the Polaris submarines under construction for the US Navy, and which was cancelled as a result of the defence policy decisions of 1959–60.

'Golf II' and 'Hotel II'

Retro-fitting the existing missile submarines of the 'Golf' and 'Hotel' classes was accompanied by a number of other modifications, at least in the case of the former.

Refitted 'Golfs', which subsequently received the NATO designation 'Golf II', were given the now standard sonar outfit comprising the Feniks passive array, housed in a bow bulge with a prominent wrap-around 'window', plus the Herkules active sonar already fitted in the early boats of the class. They were also fitted with the tall hinged mast carrying a whip aerial for HF communications which was a feature of other Soviet submarines of the period. As with the 'Hotel', the mast was located to starboard close to the forward end of the conning tower, and could be folded back into a recess in the deck casing.

Subsequently, during the early 1970s, a number of 'Golf II's were fitted with towed buoys for Very Low Frequency (VLF) communications to enable them to receive instructions when operating fully submerged. Because of the lack of space between the inner and outer hulls, special housings for the buoys were constructed atop the deck casing. On some units the housing was built as an extension of the after end of the fin, while in others it was located close to the stern (*see* drawings). It is perhaps significant that this later modification was restricted to the 'Golf', which was specifically excluded from the terms of the SALT I Treaty. The 'Hotel' class, which had nuclear propulsion, was included in the SALT Treaty provisions, and would therefore be withdrawn from service during the

Above: *Since the early 1970s several units of the 'Golf II' class have received towed buoys for VLF communications. These two photographs are of the variant that has the buoy housing atop the after deck casing (MoD/US Navy)*

1970s to enable the Soviet Navy to complete SSBNs of more modern design. The 'Hotels' were not therefore retro-fitted for VLF communications.

Thirteen 'Golfs' of the original total of 23 were converted to the 'Golf II' configuration, the conversions being carried out at Severodvinsk (eight units) and Komsomolsk (five units) from 1963, the year in which the SS-N-5 missile entered service, until 1968. Three of the unmodified boats were subsequently converted to serve as trials ships for new ballistic missiles, and a further three to a communications (SSQ) configuration (*see* Chapter 8), but the remaining 'Golf I's were to be steadily phased out during the 1970s.

All eight units of the 'Hotel' class underwent conversion to fire the SS-N-5 missile, and subsequently received the designation 'Hotel II'. The conversions were carried out at Severodvinsk from 1963 to 1967, and some sources state that the last of the original boats was completed to the new configuration. Apart from modifications related to the launch tubes and the missile fire control system, the 'Hotel II's appear to have been little altered. One boat was subsequently converted to serve as a missile trials ship in 1969–70, and a further unit was modified as a communications submarine (SSQN) during the late 1970s.

Cruise missile submarines

The defence policy decisions of 1959–60 affected not only the Soviet Navy's ballistic missile programme, but

Above and right: *These two units of the 'Golf II' class have the alternative VLF buoy arrangement, in which the buoy housing is built out as an extension of the conning tower. The twin doors atop the housing can be made out in the overhead view. (Both US Navy, 1977)*

also the programme of strategic cruise missile submarines which had just begun.

This is perhaps more surprising, given that the SS-N-3 missile had a maximum range considerably below 1,000 kilometres, and therefore did not qualify for absorption within the newly formed Strategic Rocket Forces. Nevertheless, the last of the five 'Echo I's was laid down at about this time, and the 'Whiskey Long Bin' conversion programme ground to a halt shortly afterwards, with only six or seven boats converted out of a projected total of 72.

The SS-N-3C Shaddock undoubtedly paid the penalty for being a strategic, rather than an operational-tactical missile. The 1,000km figure appears to have been established as a convenient criterion to differentiate between strategic land-based and tactical land-based missiles in order to determine which missiles should be controlled by the Strategic Rocket Forces, and which should remain under the control of the Ground Forces.

As far as the Soviet defence policy-making body was concerned, it would appear that this distinction was not applicable to sea-launched missiles. Range was not a key factor, as the missiles could be transported within range of their targets. In this sense, the SS-N-3C was just as much an 'inter-continental' missile as the SS-8. Moreover, there was no way that the inertially-guided SS-N-3C could be employed to further the tactical-operational objectives of the Soviet Navy. It therefore had no future outside the framework for a

Left: *Overhead of a submarine of the 'Hotel II' class taken in 1975. The recess for the hinged HF communications mast can be clearly seen running alongside the conning tower. The latter is more streamlined than the conning tower of the 'Golf', with all hard edges eliminated to improve waterflow. (MoD)*

coherent nuclear strategy established by the policy committees. That strategy was now to be based on land-based intercontinental ballistic missiles. The SS-N-3C was therefore abandoned in favour of the latter in much the same way that Regulus was abandoned by the US Navy in favour of Polaris.

The parallels with the American experience were to continue. Beginning in 1968 – only six years after the last boats of the class had been completed – the five 'Echo I's were taken in hand at Komsomolsk for the removal of their missile launchers. They were subsequently converted to an SSN configuration, and emerged from refit with a remodelled bow based on that of the 'Foxtrot' diesel boats, with a Feniks passive array topped by an above-bow dome for an active sonar. Apart from the installation of this somewhat dated sensor outfit they remained largely unmodified, and they were to serve in this configuration well into the 1980s, despite frequent breakdowns in their notoriously unreliable propulsion machinery.

The dozen or so SS-N-3C-equipped medium submarines of the 'Whiskey Twin Cylinder' and the 'Whiskey Long Bin' classes remained in that configuration for the rest of their service lives. Clearly it was not worth converting them back to conventional attack submarines, as it was their limited value in the latter role that had been partly responsible for their conversion. Given their relatively short endurance, they represented less of a threat to the newly formed Strategic Rocket Forces than did the nuclear-powered 'Echo'. In

Below: *The drawing depicts the 'Echo I' in its SSN configuration.*

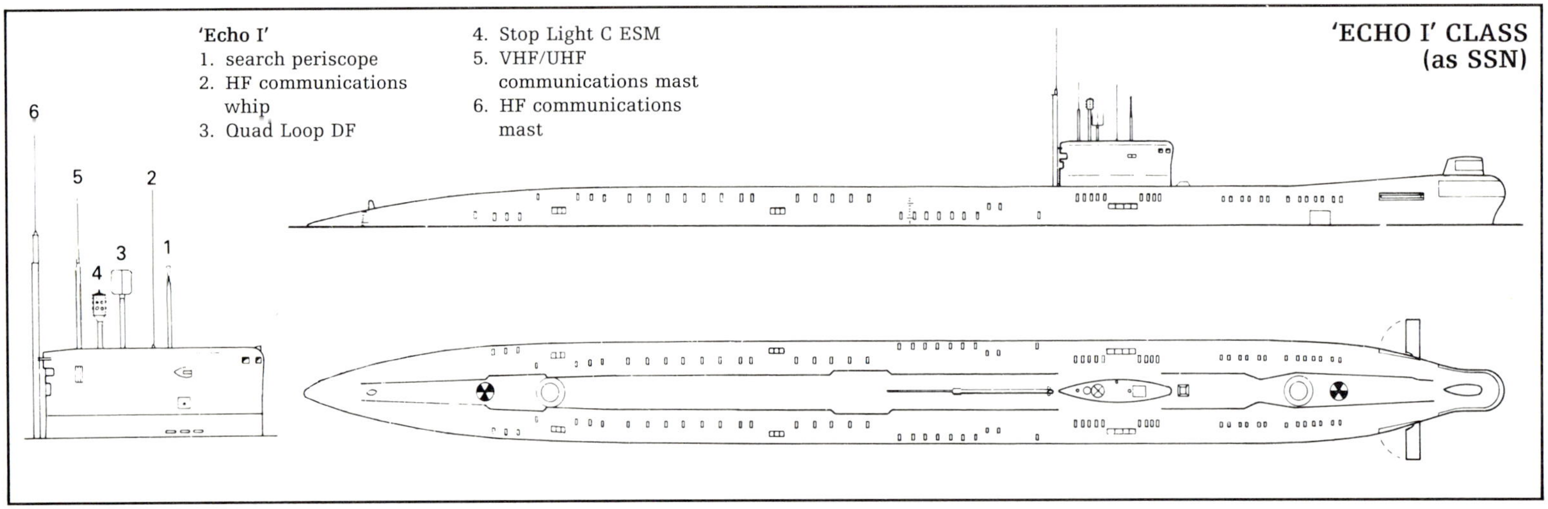

'ECHO I' CLASS

Soviet designation: Podvodnaya Lodka Atomnaya (PLA) (Submarine, Nuclear-Powered)

Conversion:
Converted at Komsomolsk-na-Amur 1968–75.

Displacement:
Surfaced: 4,600 tons
Submerged: 5,400 tons

Dimensions:
Length: 110m (361ft)
Beam: 9m (30ft)
Draught: 7.5m (25ft)

Propulsion:
Two pressurized water-cooled reactors driving geared steam turbines; two shafts with 5-bladed propellers; 30,000shp for 25 knots max.

Complement 75

Weapons:
Six bow tubes of 533mm (21in) diameter for anti-ship torpedoes; four stern tubes of 400mm (16in) diameter for anti-submarine torpedoes; maximum capacity eighteen 533mm torpedoes, eight 400mm torpedoes.

Sensors:
Feniks MF passive sonar; Herkules HF active sonar; search and attack periscopes; Snoop Tray surface surveillance radar; Stop Light ESM; Quad Loop DF.

Communications:
HF and VHF radio.

particular, there was no suggestion that they could or would be used for regular patrols off the coasts of the United States. They were therefore retained by the Soviet Navy, and employed as theatre nuclear forces.

A new mission

The Soviet submarine arm was desperately in need of a new mission. Virtually its entire construction programme was geared to producing submarines capable of strategic nuclear strike against the US mainland. That mission was now to be performed by land-based inter-continental missiles under the control of the Strategic Rocket Forces.

Fortunately for the Soviet Navy a new mission was already at hand. From 1955 the US Navy had begun to complete a new generation of 60,000-ton 'super-carriers' designed to operate long-range nuclear bombers. Four of these ships had already been completed by the time of the Soviet policy review of 1959–

Above and right: *Two views of the 'Echo I' SSN conversion. Note the reconfigured bow and the long, uncluttered hull. Following conversion these submarines served to boost SSN numbers in the Pacific at a time when the new-built 'Victor' was being allocated almost exclusively to the Northern Fleet. A series of mechanical breakdowns during the 1980s, together with the increased availability of the 'Victor', has resulted in a steady decommissioning of the 'Echo I's. (MoD/US Navy)*

Above: *Overhead of an 'Echo I' following SSN conversion during the early 1970s. The SS-N-3 missile launchers have been removed and the hull plated over. The new sonar suite is identical with that of the 'Foxtrot', with the Feniks passive array topped by an elliptical dome for the Herkules attack sonar. (US Navy, 1975)*

60, and a further three, including the nuclear-powered *Enterprise*, were due to enter service within the next two years.

The A-3 Skywarrior, a large jet-propelled bomber with a strike radius of about 1,200 nautical miles, could threaten the cities of the Soviet Union from launch points outside Soviet sea-space. These carriers could not therefore be countered by the traditional Soviet anti-surface tactic of employing large numbers of small surface ships and medium submarines, operating under a friendly air umbrella.

The only effective counter to carriers operating in the Norwegian Sea or the north-west Pacific was a combination of long-range maritime bombers and submarines. Both would have to launch their weapons from outside the ring of task force defences if they were to be effective. The US Navy had by now perfected the art of carrier air defence, employing sophisticated airborne early-warning (AEW) aircraft to detect intruders at long range and to vector fighter combat air patrols out to intercept them.

Anti-submarine defence was also proceeding by leaps and bounds, with the advent of large, powerful low-frequency sonars on US surface escorts, complemented by the ASROC anti-submarine missile. The Soviet submarine arm had no experience of attacking large formations of ships. Its only surface targets during the Second World War had been isolated German warships, or small convoys of lightly escorted merchant ships. It was by no means certain that Soviet submarines, even if skilfully handled, would be able to penetrate a protective screen of surface escorts equipped with modern acoustic detection devices and armed with missiles and homing torpedoes. The US Navy had long experience of anti-submarine operations and the Soviets, with their predilection for historical example, cannot have failed to note the success of these operations during the latter part of the Pacific War, when not a single Japanese submarine managed to penetrate task force defences.

The answer to both the air defence and the anti-submarine defence problems was a stand-off missile, which would enable both the maritime bomber and the submarine to launch their weapons without first having to penetrate the defensive rings thrown up around the carriers.

The first implication of a stand-off missile was that it would have to be large in order to have the necessary range to be fired from long distances, and to be able to carry either a nuclear warhead or a high-explosive payload sufficiently powerful to disable a carrier. It would also have to have a relatively high flight profile.

The sea-skimming mode subsequently adopted for some Western anti-ship missiles, notably Exocet, appears not to have been considered.

A nuclear warhead, which was almost certainly the payload initially envisaged for such a missile, would be most effective if exploded directly above the task force. Moreover, a sea-skimming flight profile would be of little value even if a conventional payload were used in conjunction with an active homing radar, because the tendency of such a missile is to attack the first target it encounters. If fired against a task force, the missile would most likely home on an escort ship in the outer screen.

A missile approaching the task force at high altitude, on the other hand, would be able to look down on the task force, and its homing radar would therefore 'see' all the available targets. The missile would then home on the largest radar 'blip' which, it was to be hoped, would be returned by the carrier at the centre of the task force.

In order to extend the radar horizon of the submarine, specially fitted targeting aircraft would be employed. These would provide the shore control post with accurate data on the position, heading, and composition of the enemy task force, and would be able to transmit a radar 'picture' of the force direct to the submarine via a video data link, to enable it to make any necessary mid-course corrections to the flight of the missile.

The 'Echo II'

The SS-N-3 Shaddock missile, already available as an inertially guided weapon for strategic nuclear strike, was an eminently suitable vehicle for conversion to the anti-ship mode. It had the necessary range, could carry a 1,000kg payload plus a homing radar, and was susceptible to guidance.

The development of a carrier submarine was a relatively easy matter. The 'Echo I' was already in production as a launch vehicle for the SS-N-3C inertially guided missile. The design would simply require modification to incorporate the necessary facilities for post-launch guidance to ensure that the missile was directed towards the last known position of the task force.

Interestingly, the battery of six paired SS-N-3 launchers installed in the 'Echo I' was deemed insufficient, and a further pair of launchers was added abeam the after part of the conning tower. This modification can only be satisfactorily explained by the adoption of a conventional (HE) payload for the SS-N-3 missile as an alternative to a nuclear warhead at an early stage in the submarine's design process. Six missiles would have been more than sufficient against a carrier task

Below: *The drawings depict the 'Echo II' as built.*

'ECHO II'

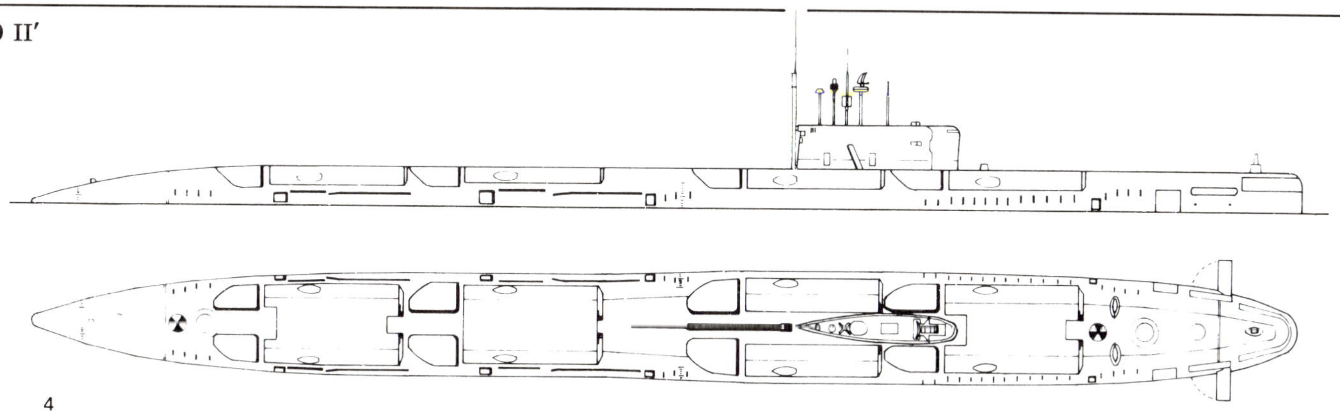

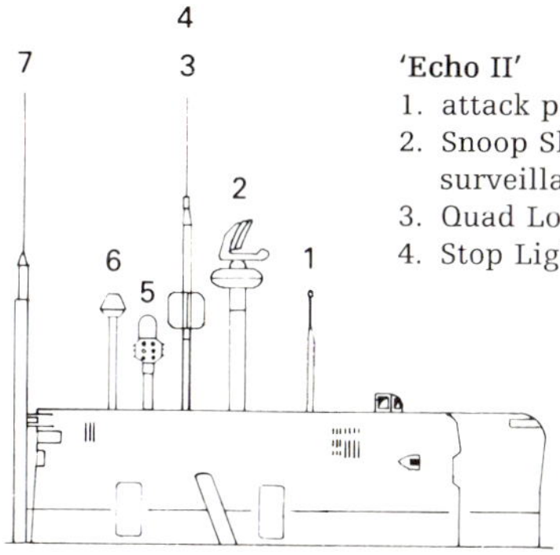

'Echo II'
1. attack periscope
2. Snoop Slab air surveillance radar
3. Quad Loop DF
4. Stop Light ESM
5. VHF/UHF communications mast
6. missile data link?
7. HF communications mast

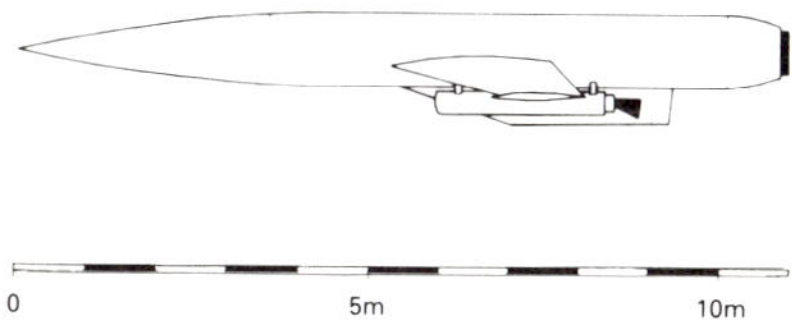

SS-N-3A SHADDOCK

Length:	10.2m
Span:	5m
Diameter:	975mm
Propulsion:	turbojet + 2 solid-fuel boosters
Guidance:	command/active radar homing
Range:	250nm (460km)
Warhead:	1,000kg HE or nuclear
In Service:	1962

Left: *The four pairs of launch tubes for the SS-N-3 Shaddock missile can be clearly seen in this bow quarter view. Note the prominent indentations behind each pair of launchers which serve to deflect the blast when the missiles are launched. (MoD, 1978)*

Right: *An 'Echo II' with its distinctive Snoop Slab air surveillance radar in the raised position. There are individual variations in the sonar outfit of the 'Echo IIs'. The unit in the photograph, taken in 1984, retains the sonar dome atop the bow. (MoD)*

SS-N-3 LAUNCH SYSTEM

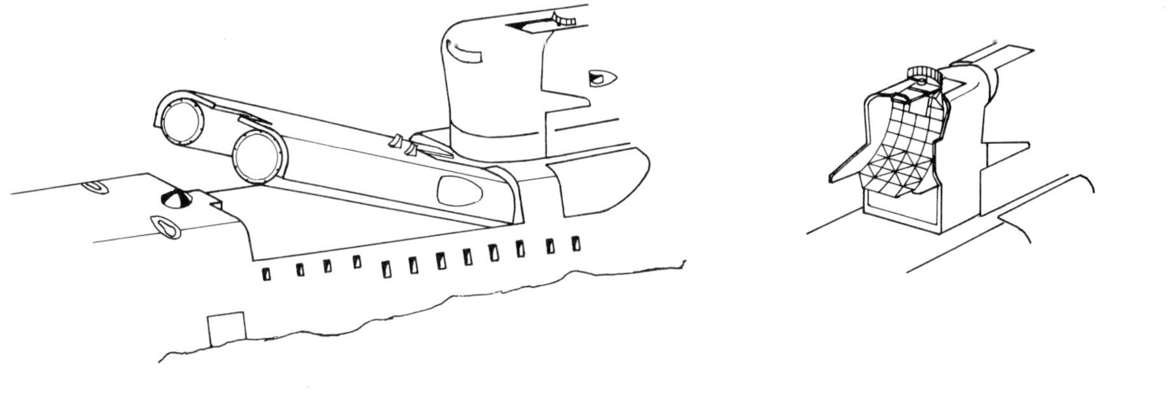

Left: *The drawings show the SS-N-3 missile launchers in the raised position for firing, and the forward section of the conning tower rotated to expose the Front Door/Front Piece missile tracking radar*

'ECHO II' CLASS

Soviet designation: Podvodnaya Lodka Atomnaya Raketnaya Krylataya (PLARK) (Cruise Missile Submarine, Nuclear-Powered)

Construction:
Built at Shipyard No. 402, Severodvinsk; Komsomolsk-na-Amur.
29 units completed 1962–7: two in 1962, four in 1963, seven in 1964, seven in 1965, six in 1966, three in 1967.

Displacement:
Surfaced: 5,200 tons
Submerged: 6,200 tons

Dimensions:
Length: 116m (380ft)
Beam: 9.8m (32ft)
Draught: 7.5m (25ft)

Propulsion:
Two pressurized water-cooled reactors driving geared steam turbines; two shafts with 4-bladed propellers; 30,000shp for 23 knots max.

Complement 100

Weapons:
Eight SS-N-3A Shaddock anti-ship missiles in paired elevating launchers; six bow tubes of 533mm (21in) diameter for anti-ship torpedoes; four stern tubes of 400mm (16in) diameter for anti-submarine torpedoes; maximum capacity eighteen 533mm torpedoes, eight 400mm torpedoes.

Sensors:
Feniks MF passive sonar; Herkules HF active sonar in some units; search and attack periscopes; Front Door/Front Piece missile guidance radar; Snoop Slab air surveillance radar; Stop Light ESM; Quad Loop DF.

Communications:
HF and VHF radio.

force had nuclear warheads been used. However, conventional warheads would only be fully effective if saturation missile strikes could be launched in order to overwhelm task force defences. Eight missiles were needed if there were to be a satisfactory 'mix' of conventional and nuclear missiles aboard each submarine to cope with either eventuality.

The additional pair of launchers brought with them an increase in length of some six metres and displacement rose by 600 tons. Given that the propulsion plant remained the same, this must have resulted in a slight loss of speed, but high tactical speed was less important in this class than it was in the 'November' because of the long range of the SS-N-3 missile, which made it a relatively easy matter to reach a satisfactory firing position in advance of the task force.

The major external difference between the 'Echo I' and the 'Echo II' was the conning tower, which was modified to accommodate the missile guidance radars. Although identical in height, the conning tower of the 'Echo II' was some five metres longer. The forward section housed the Front Door/Front Piece radar combination, which faced aft when stowed. Once the

submarine had surfaced to prepare for missile launch, the section containing the radar was rotated through 180 degrees (*see* drawing). The radar was then free to elevate to the required angle to track the missile.

Front Door, which is the larger element in the antenna combination, is essentially a 'nodding' height-finder of conventional parabolic lattice configuration. It is paired with a smaller parbolic antenna incorporating a prominent overhead feed horn which is designated Front Piece, and which provides coverage in the horizontal plane. The two antennae combine to give missile tracking in three dimensions.

The 'Echo II' also featured a large complement of mast-mounted sensors, including some which are clearly related to the specialist role of these submarines. As with earlier types there were the standard search and attack periscopes, a Quad Loop DF aerial, and mast-mounted whip aerials for VHF/UHF and HF communications, the latter atop a hinged mast attached to the end of the fin as in the 'Echo I'. However, the Stop Light ESM antenna was a new model, with a broad, 'studded' base topped by a small radome (NATO designation: Stop Light C).

The major new antenna was the Snoop Slab radar, which was fitted in place of the Snoop Plate/Snoop Tray series of surveillance radars. Snoop Slab was significantly larger than earlier Soviet submarine radars, and there was a prominent circular base plate beneath the parabolic antenna (the 'Slab' element in the NATO designation). The large size of the antenna suggests that its function was not surface surveillance, as in the case of Snoop Plate and Snoop Tray, but air surveillance. It may have been provided as a back-up to the Front Door/Front Piece missile guidance radars, but it does not appear to have had a specific missile-related function, and may simply have been provided to give the submarines a long-range detection capability against aircraft.

Launch preparations for the SS-N-3 missiles are estimated to take at least 20–25 minutes, and the submarine would have to remain on the surface throughout this time. Preparations would include rotating the forward section of the conning tower to expose the missile guidance radar, elevating the massive cylindrical launchers, opening the doors at each end of the launcher, and making the customary checks on the missile. The submarine would also have to remain on the surface after launch for a further twenty minutes or so, in order to track the missiles and provide any necessary course corrections.

Above: *An 'Echo II' running on the surface in northern waters. (MoD 1986)*

The Shaddock SS-N-3A variant which equips the 'Echo II' has a maximum range estimated at 200 nautical miles in the surface attack role. The attack aircraft that operated from US aircraft carriers during this period had a strike radius more than double this figure. Although the attack aircraft would, like the 'Echo II' missiles, require launch preparation and would have to make a lengthy transit at subsonic speed, there was clearly a period of vulnerability when the submarine might be subject to attack on the surface. The Snoop Slab radar appears to have been provided against such an eventuality, as long-range detection of hostile aircraft would give the submarine time to make ready for diving.

The other new antenna first seen aboard the 'Echo II' was a small mast-mounted radome. This antenna has not reappeared on any other class of Soviet submarine, and is therefore almost certainly related to missile guidance. It may be the aerial which receives the video data link from the targeting aircraft.

The 'Juliett'

The first unit of the 'Echo II' class was probably laid down at the Komsomolsk Shipyard in 1960. Indeed, the earliest boats may have been laid down as 'Echo I's, and modified while on the stocks. Following the completion of the last boats of the 'Echo I' class, the 'Echo II' entered full production at a rate of approximately two per year.

A second production line was opened at Severodvinsk, which was due to complete its last 'Hotels' in 1962–3. Additional construction berths were also becoming available with the phasing-out of 'November' construction, so that by the mid-1960s the Severodvinsk Shipyard was completing between four and five 'Echo II's per year. The 'Echo II' was the only Soviet nuclear-powered submarine in full production over this period.

Twenty-nine units were built in all – eighteen at Severodvinsk and eleven at Komsomolsk. Deployment initially followed a similar pattern, but some Arctic-built units were subsequently transferred to the Pacific as more modern submarines armed with anti-ship missiles became available to the Northern Fleet.

The importance of the anti-carrier programme was such that even these numbers were deemed inadequate. Moreover, the Soviets had still not resolved the problem of what to do with the construction capacity made available by the abandonment of the medium submarine building programme. The solution to both problems was to build a diesel-powered counterpart to the 'Echo II', and to allocate construction to the Gorky Shipyard, which was at that time engaged in fitting out the last of the 'Romeos'.

The 'Juliett', unlike other Soviet diesel-powered

Right: *The 'Juliett' is the diesel-powered counterpart of the 'Echo II'. The relatively small size of the submarine necessitated a reduction from four to two pairs of missile launchers. (US Navy, 1986)*

missile boats of the period, was a completely new design. Marginally shorter than the 'Foxtrot', it had a substantially broader outer hull casing to accommodate launch tubes for the SS-N-3A Shaddock missile. There were two pairs of launchers – one immediately forward of the conning tower, one aft – of identical configuration to those installed in the 'Echo II', with the same indentations for blast deflection in the outer casing.

The propulsion system was based on a more powerful diesel rated at about 2,500bhp. Earlier diesel/electric propulsion systems had all been based on the standard post-war 2,000bhp Type 37D model, which had powered two shafts in the medium 'Whiskey' and 'Romeo', and three in the larger boats. The adoption of the more powerful diesel for the 'Juliett' enabled the Soviets to revert to a more economical two-shaft arrangement. A large diesel also facilitated the installation of more powerful electric motors on each of the two shafts.

The torpedo armament of the 'Juliett' is thought to be identical with that of the 'Echo II', with six tubes of 533mm diameter forward and four short 400mm stern tubes for anti-submarine homing torpedoes. The sonar outfit comprises the standard Feniks/Herkules combination, with the latter in an above-bow sonar dome.

The outfit of sensor masts is also identical with that of the 'Echo II', with two important exceptions. The HF whip aerial is carried not atop a separate hinged mast, as on the 'Echo', but on a fully-retractable mast housed in the after (raised) section of the conning tower. Snoop Slab is fitted in place of Snoop Tray, as in the 'Echo', but there is no sign of the small unidentified mast-mounted radome seen aboard the nuclear boats.

The forward end of the conning tower, which houses the Front Door/Front Piece guidance radar, is identical with that of the 'Echo', but the conning tower itself is some three metres longer in order to accommodate a snorkel and the diesel exhaust.

Sixteen 'Julietts' were completed at the Gorky shipyard between 1961 and 1969. Some sources have suggested that this programme was originally intended to be much larger,[1] but the available evidence contradicts this view. The span of the 'Juliett' programme provides an almost exact parallel with that of its nuclear-powered counterpart, the 'Echo II'. Not only did construction take place over a period of about nine years, but the production rate throughout that period was a steady two per year, which suggests that the class was never intended for mass production. More-

Right, top and bottom: *The hull-form of the 'Juliett' is broader than that of the 'Foxtrot'. The exceptionally long conning tower houses not only the standard outfit of sensor masts, but also the Front Door/Front Piece missile guidance radar. As in the 'Echo II', the front section rotates prior to missile launch. (US Navy, 1987/MoD, 1979)*

'JULIETT' CLASS

Soviet designation: Podvodnaya Lodka Raketnaya Krylataya (PLRK) (Cruise Missile Submarine, Diesel-Powered)

Construction:
Built at Krasnaya Sormovo Shipyard, Gorky.
Sixteen units completed 1961–9: one in 1961, two in 1962, two in 1963, two in 1964, two in 1965, two in 1966, two in 1967, two in 1968, one in 1969.

Displacement:
Surfaced: 3,000 tons
Submerged: 3,750 tons

Dimensions:
Length: 90m (295ft)
Beam: 10m (33ft)
Draught: 7m (23ft)

Propulsion:
Diesel/electric drive on two shafts; two diesels, 5,000bhp for 12 knots surfaced; two electric motors, 3,500hp for 12 knots submerged.

Complement 80

Weapons:
Four SS-N-3A Shaddock anti-ship missiles in paired elevating launchers; six bow tubes of 533mm (21in) diameter for anti-ship torpedoes; four stern tubes of 400mm (16in) diameter for anti-submarine torpedoes; maximum capacity twelve 533mm torpedoes, eight 400mm torpedoes.

Sensors:
Feniks MF passive sonar; Herkules HF active sonar; search and attack periscopes; Front Door/Front Piece missile guidance radar; Snoop Slab air surveillance radar; Stop Light ESM; Quad Loop DF.

Communications:
HF and VHF radio.

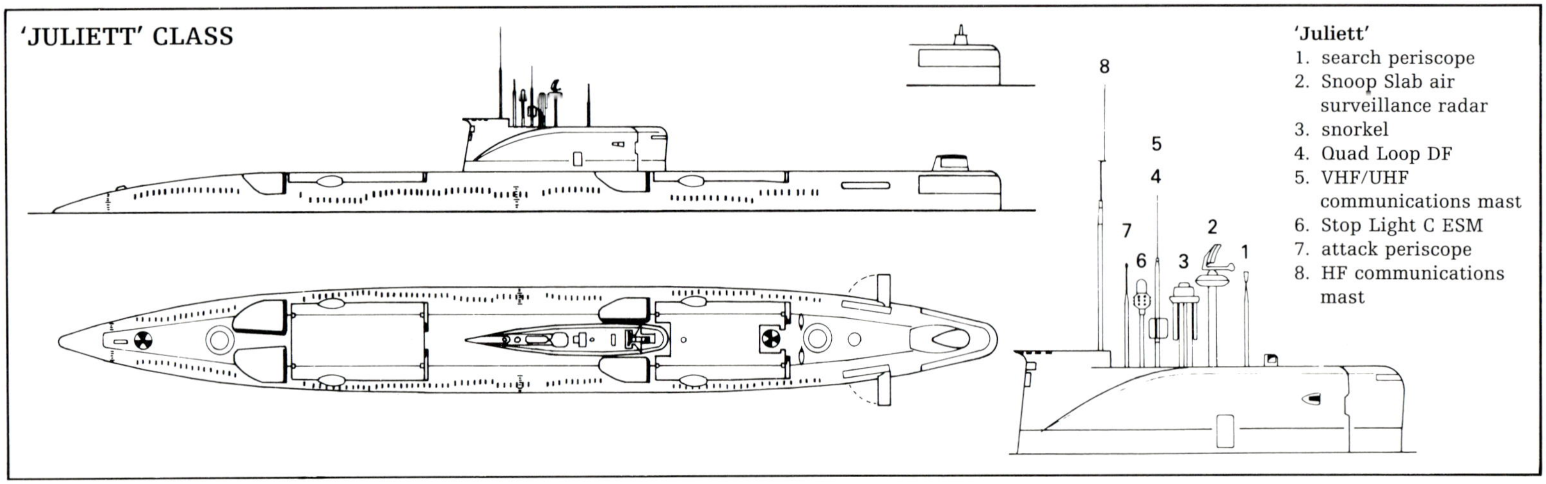

Left: *The 'Whiskey Canvas Bag' was to have been employed as part of a picket line to detect incoming strategic bombers. Only five or six units are thought to have undergone conversion. (US Navy, 1970)*

over, two units per year was probably close to the total capacity of the Gorky Shipyard at this time, as the shipyard was being partially rebuilt to enable it to undertake the construction of nuclear-powered submarines from the mid-1960s onwards.

The deployment pattern of the 'Juliett' was somewhat different from that of the 'Echo II'. Two-thirds of the class operated with the Northern Fleet until the late 1970s, while only five or six boats served with the Pacific Fleet. This imbalance is perhaps not surprising, given that all sixteen units were built in the western USSR.

The 'Whiskey Canvas Bag'

The period that followed the major policy changes of 1959–60 also witnessed the emergence of yet another 'Whiskey' conversion.

The 'Whiskey Canvas Bag', as it subsequently became known, was a less extensive modification of the original 'Whiskey' design than the 'Long Bin'. Conversion work therefore proceeded in parallel, even though in conceptual terms the 'Canvas Bag' post-dates the strategic missile conversion.

Modification involved extending the conning tower to twice its original length, and incorporating a large air surveillance antenna, complete with its turning mechanism, in the after section. The forward end of the conning tower remained virtually unchanged, except that the Quad Loop DF aerial, which had previously

'WHISKEY CANVAS BAG' CLASS

Soviet designation: Podvodnaya Lodka Vozdushnogo Nablyudeniya (?) (PLVN) (Radar Picket Submarine) (PL) (Submarine)

Conversion:
5–6 units converted 1960–3.

Displacement:
Surfaced: 1,080 tons
Submerged: 1,450 tons

Dimensions:
Length: 76m (249ft)
Beam: 6.3m (21ft)
Draught: 4.8m (16ft)

Propulsion:
Diesel/electric drive on two shafts; two Type 37D diesels, 4,000bhp for 17 knots surfaced; two electric motors, 2,500hp for 13.5 knots submerged.

Complement 55

Weapons:
Four bow tubes of 533mm (21in) diameter for anti-ship torpedoes; maximum capacity eight torpedoes.

Sensors:
HF passive sonar; Tamir 5 HF active sonar; Boat Sail long-range air surveillance radar; search and attack periscopes; Snoop Plate surface surveillance radar; Stop Light ESM; Quad Loop DF.

Communications:
HF and VHF radio.

'WHISKEY CANVAS BAG' CLASS

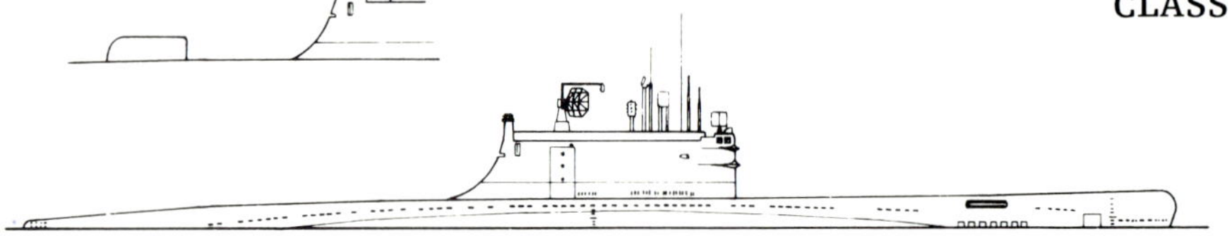

Above: *Close-up of the elongated conning tower, topped by the large Boat Sail air surveillance antenna. The turning mechanism for the antenna is housed within the conning tower itself. The Boat Sail antenna is collapsible, and is folded and covered when the submarine dives. Both the units depicted retain the original 'gun deck' abaft the conning tower. This suggests that the submarines that underwent conversion were early units of the class. (US Navy, 1970)*

Left: *At least one unit of the 'Whiskey Canvas Bag' sub-group appears to have been modified as a communications submarine during the mid-1970s* (see *upper drawing). The conversion involved the removal of the Boat Sail air surveillance radar, and its replacement by a small satellite dish. The housing abaft the conning tower is almost certainly for a VLF communications buoy.*

been fixed to the 'chimney' for the diesel exhaust, was relocated forward above the bridge windows.

The air surveillance radar, which subsequently received the NATO designation Boat Sail, comprised a parabolic lattice antenna with a prominent feed horn suspended above it. The antenna was 'collapsible', and was deployed only when the submarine was running on the surface. When folded it was generally concealed beneath a protective covering which gave rise to the designation 'Canvas Bag'.

Bulges in the sides of the conning tower immediately beneath the radar antenna housed the turning mechanism, and beneath the turning mechanism, within the hull, there was presumably a radar office located directly abaft the original control room. The after torpedo tubes were apparently removed, and the after torpedo room converted to accommodation spaces.

The concept of the radar picket submarine was by no means new. The idea had first been evolved by the US Navy in the final year of the Second World War as an answer to the high attrition rates of radar picket destroyers stationed ahead of the carrier task forces to provide early detection of hostile air attacks. Heavy losses sustained at the hands of Japanese *Kamikaze* pilots led to the suggestion that this mission be performed by submarines, which would operate on (or close to) the surface to provide initial detection, but could then dive to avoid attack.

The concept was developed during the 1950s, with some warbuilt 'fleet' submarines undergoing conversions that equipped them not only with air surveillance radars, but with height-finders and fighter control centres to enable them to vector out the carrier combat air patrols (CAPs) to intercept hostile aircraft. Two purpose-built diesel boats, *Sailfish* (SSR-572) and *Salmon* (SSR-573), were also completed during this period.

The principal difficulty with the concept was that conventionally powered submarines had inadequate speed on the surface to keep pace with a carrier task force. The final development was therefore the design of a large submarine powered by two nuclear reactors, the *Triton* (SSRN-586). The US radar picket programme did not survive beyond the 1950s, however, and *Triton* reverted to attack submarine status in 1961.

Although in technical terms the 'Whiskey Canvas Bag' bears a passing resemblance to the US Navy's radar picket submarines, the rationale for the conversion was very different.

Like the 'Echo II' and the 'Juliett', the Soviet radar pickets were a response to the threat posed by the US carriers and their long-range nuclear bombers. Their function was essentially strategic, not tactical (as was the case with the US radar pickets). They were equipped neither with height-finders, nor with control centres capable of directing friendly fighter aircraft. They would simply be placed ahead of the expected track of a carrier task force, and would transmit data concerning the number and heading of enemy aircraft to a shore control post, which would then analyse this

information, together with data from other sources, in order to achieve the best disposition of fighter resources to intercept the enemy force.

Only five (possibly six) 'Whiskey Canvas Bag' conversions were completed before the Soviet radar picket programme was abandoned, probably in favour of more sophisticated aerial reconnaissance and early warning systems.

The targeting problem

The development by the Soviets of a long-range cruise missile for anti-ship attack could be realized only within the framework of Combined Arms tactical thinking. Targeting was a major problem which would have to be resolved by creating an elaborate network of off-board sensors and the establishment of rigidly controlled operating procedures.

The speed of the SS-N-3 missile is generally estimated at about Mach 0.9 – i.e., no greater than that of a subsonic carrier-borne attack aircraft. For long-range strike this means that there would be a substantial 'dead time' between the launch of the missile from the submarine and its arrival in the vicinity of the target. This interval could be exploited by the target, which could engage in evasive manoeuvres designed to disrupt potential enemy strikes.

Above: *The Tu-95 Bear D is a key element in the reconnaissance/targeting system on which submarines armed with the SS-N-3 anti-ship missile depend. Note the large underbelly fairing for the Big Bulge surveillance radar. (US Navy, 1980)*

Above: *Of the 'Juliett' and the 'Echo II', only the latter has sufficient speed to be able to close a hostile task force from the flank. The 'Echo' is a noisy boat at speed, and such a manoeuvre would expose it to early detection. (US Navy, 1970)*

The missile submarine is regarded by the Soviets essentially as a weapons carrier, and is not equipped to provide initial detection of an enemy task force. Indeed, if it were close enough to detect and classify the enemy force it would be so far within its own missile envelope that it would be unsafe to surface for an attack. Initial detection of the enemy force prior to the establishment of reconnaissance satellite networks during the late 1970s would almost certainly have come from widely spaced, high-frequency direction-finding stations, using triangulation to obtain the range and bearing of enemy radio transmissions. This would have been followed by the dispatch of a pair of specially fitted Bear D maritime reconnaissance aircraft, bristling with electronics aerials, to locate and classify the target. Once the target had been located, the course, bearing and approximate composition of the force would be relayed back to a shore-based command post. The shore-based commander would then look at the disposition (and availability) of his own forces, and would attempt to position those forces in such a way as to ensure a co-ordinated strike at a predetermined point.

The missile submarine might be some distance from this predetermined launch point, and would either have to be positioned ahead of the task force, or would have to make a high-speed transit from the beam or from the stern quarter. Probably only the nuclear-powered 'Echo' would be capable of making such a transit; the diesel-powered 'Juliett', with a maximum speed on the surface of only 15–16 knots even in a flat sea, would find it difficult to close a 25-knot task force from any angle other than straight ahead. Both would require constant updating of the target's course and position.

As we have seen, between 20 and 30 minutes would probably be required for missile preparation. During this period, speed would have to be reduced, probably to no more than 8 knots. Further updates of the target's position would again be required if this period were not to be added to the 'dead time' which would occur during the flight of the missile. These updates would have to come either from the shore control post or direct from the reconnaissance aircraft, assuming these were still in contact with the target. If from the latter, the submarine would need to know the precise position of the aircraft itself in relation to the enemy task force. A heavy reliance would therefore have to be placed on radio communications and data links. Presumably one function of the long-range Snoop Slab air surveillance radar installed in both the 'Echo' and the 'Juliett' is to track friendly targeting aircraft.

The missiles are then launched on command from the shore control post. The Front Door/Front Piece missile tracking radars enable the submarine to provide guidance and course correction during the early phase of the missile's flight, but assistance from the Bear D reconnaissance aircraft is required for over-the-horizon targeting. It is reported that the radar picture from the Big Bulge antenna of the latter can be transmitted directly to the submarine via a video data link, thereby enabling the submarine to make the necessary mid-course corrections.

The importance of constant updating of the targeting information throughout these complex operations can readily be seen. If detailed aerial reconnaissance were frustrated by the carrier fighter combat air patrols (CAPs), the shore control position would have to rely on relatively infrequent HF/DF 'cuts' and intelligent guesswork to predict the position of the task force at a given moment. Even if continuous aerial reconnaissance could be maintained, there would be inevitable delays in the transmission and reception of targeting data and instructions between the various missile and reconnaissance platforms, combined with difficulties in determining the precise position of those platforms in a constantly changing maritime scenario.

A missile flying at Mach 0.9 over a distance of 200 nautical miles would take some 20 minutes to arrive over its target. Failure to provide an update of target

Left: *Co-operation between the Soviet missile submarine and surface units is considered desirable in order to achieve a co-ordinated missile strike. Here a 'Juliett' is seen in company with a Rocket Cruiser of the 'Kresta I' class. (US Navy)*

Above: *The 'Echo II' has to spend a considerable time on the surface both prior to and following missile launch.*

co-ordinates during the preparations for missile launch would add a further 20–30 minutes 'dead time'. If transit were required to the predicted launch point, further time might elapse.

In 20 minutes a 25-knot task force would have travelled about 8 nautical miles; in 50 minutes it would have moved 20nm; in an hour 25nm, and so on. Its course would not necessarily be constant, and could involve 90- or 180-degree turns which might be related to standard carrier air operations, but which might equally be undertaken as evasive manoeuvres. This means that the SS-N-3 missile might well arrive at the estimated target position and find nothing but an expanse of open sea. In theory, Soviet cruise missile submarine operations as conceived during this period were logical and 'scientific'. In practice, the number of variables and uncertainties made it highly unlikely that the ideal conditions for a co-ordinated missile strike against a hostile carrier task force could be achieved.

The defence of Soviet sea-space

Soviet submarines of the first post-Stalin programme had a primary strategic mission, involving long-distance operations against the land mass of the United States. They were therefore characterized by large size and long endurance, with nuclear power being favoured for propulsion.

The submarines of the anti-carrier programme were of similar size, and the 'Echo II' was nuclear-powered, but the reasons were different. Large size was necessary to accommodate a main battery of SS-N-3 Shaddock missiles, each of which had a length of 10.2 metres. Even at close to 90 metres in length, the diesel-powered 'Juliett' could carry only four missiles – half the number embarked in the 'Echo II'. This figure was barely adequate, given the need for a conventional/nuclear warhead mix and saturation missile strikes. Such strikes were necessary in order to overcome enemy air defences and to compensate inherent inaccuracies in the targeting and guidance systems. It therefore seems likely that the 'Julietts' were intended to operate in pairs, the pair being counted as a single unit.

In theory nuclear power gave the 'Echo II's long range, and it was initially feared in the West that they would threaten surface vessels operating in the open oceans. However, the reliance of these submarines on an elaborate network of aerial reconnaissance for target detection and classification, and on the presence in the target area of specially equipped missile target-

'ECHO II' CLASS

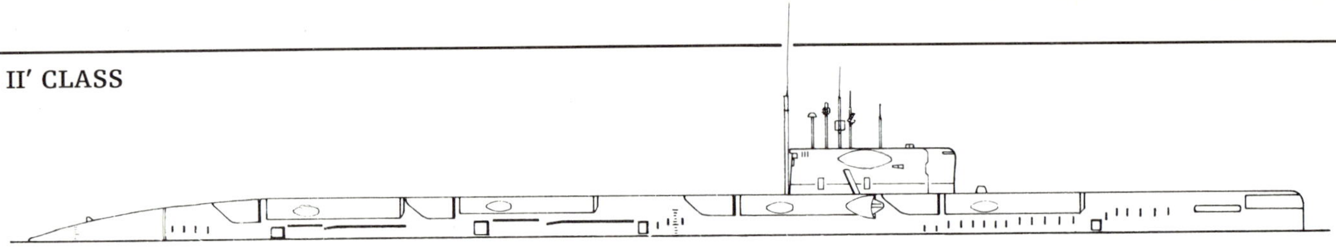

Left: *Profile drawing of an 'Echo II' modified to fire the SS-N-12 missile. Note the revised outfit of sensor masts.*

There can be little doubt that the most favourable interception course for one of these submarines would be from ahead. Preparations for missile launch could then be carried out on the surface long before the task force entered missile range, thereby safeguarding the ing aircraft, virtually ensured that they would not be thus employed. The reconnaissance and missile targeting aircraft would be most effectively employed close to their bases, as the proximity of their operating area to land would be determined by the need to re-acquire a contact initially established by HF/DF, and the need to conduct continuous reconnaissance patrols. Proximity to land would favour a high sortie rate, and long-range fighter protection might be available. Similar considerations would apply to the maritime bombers which would be expected to co-ordinate their missile strikes with those of the submarines.

The major advantage bestowed on the 'Echo II' by nuclear propulsion was the ability to approach a task force from virtually any angle once the latter had been detected. These boats could therefore be freely deployed to ensure that they were in a favourable position for early interception of an approaching task force. The diesel-powered 'Julietts', on the other hand, would be better deployed in a relatively static patrol line strung across the likely path of advance of a task force.

submarine against pre-emptive air attack. The need to reduce speed during missile preparations would make it difficult for even the nuclear-powered 'Echo' to approach from the beam. Furthermore, high-speed manoeuvres aimed at reaching a favourable launch position would risk alerting the task force to the presence of the submarine, even when running submerged, because of the high level of machinery noise emitted by these first-generation boats.

Later modifications

Beginning in the mid/late-1970s, some units of the 'Echo II' class were modified to fire the SS-N-12 Sandbox missile in place of the SS-N-3A Shaddock.

The SS-N-12 missile first entered service aboard the anti-submarine cruisers of the *Kiev* class, and has recently also been fitted in the rocket cruisers of the *Slava* class. The missile appears to be of broadly similar configuration to the SS-N-3, and is generally regarded as its lineal successor. It is about 1.5 metres longer than the SS-N-3, and maximum range is thought to be in excess of 300 nautical miles, as compared with about 200nm for the SS-N-3A/B variants of Shaddock. The SS-N-12 is an air-breather, and therefore has to be launched from the surface from launchers similar to those employed for its predecessor.

Accommodating the new missile aboard the 'Echo II' appears to have been relatively straightforward, although some minor modifications to the launchers were required. External modifications include the addition of bulged fairings to the forward end of the second pair of launchers, located abeam the fin. The precise function of these fairings is as yet unclear.

Other external modifications include prominent bulged fairings on each side of the fin, which are thought to house new satellite communications equipment related to missile targeting, and the replacement of the large Snoop Slab air surveillance radar by the standard Snoop Tray antenna. The removal of Snoop Slab may reflect Soviet concerns regarding emission control (EMCON). Whilst a long-range air surveillance radar can assist in detecting hostile aircraft it can also

Top left: *A modernized unit of the 'Echo II' class photographed in northern waters in August 1987. Note the prominent bulged fairings at the forward end of the second pair of missile launchers and on the fin. (MoD)*

Left: *The 'Echo II' continues to be an important component of the Soviet Navy's anti-surface forces, particularly in the Far East. This SS-N-12 equipped variant was photographed by a patrol aircraft of the Japanese Maritime Self-Defence Force during the early 1980s.*

serve as a 'beacon', thereby giving away the position of the submarine and enabling attack or ASW patrol aircraft to home in on its emissions.

As first completed, most 'Echo II's had a small dome for an active Herkules sonar above the bow. In all converted boats this has apparently been removed, and even some unmodified boats now lack this feature. The bow-mounted passive array has been retained.

There is some disagreement between the various sources regarding the number of 'Echo II's that have undergone conversion to fire the SS-N-12 missile. US sources claim a limited conversion programme of about six boats, while British sources favour a higher figure of ten boats converted. Whichever figure is correct, even a limited SS-N-12 conversion programme is testimony to the continuing importance of these submarines to the Soviet Navy's anti-carrier mission.

The conversion has been extended to at least one unit of the 'Juliett' class, which has received bulged fairings similar to those of the modified 'Echos' on either side of the fin. However, it appears unlikely that other units of the class will undergo conversion, as the importance of these boats appears to have been downgraded with the entry into service of the SSGNs of the 'Oscar' class in the early 1980s. Since 1981, several units have been transferred from the Northern Fleet area to the Baltic where, in addition to their anti-ship mission, they are probably tasked with theatre support of the Warsaw Pact Ground Forces. One unit is reported to have decommissioned in 1984–5.

Above: *A modified 'Juliett' photographed in the Baltic in 1982. Note the bulge for a satellite targeting reception antenna on the side of the conning tower. Since 1981 several units of the class have been transferred to the Baltic, presumably as a component of Soviet theatre nuclear forces.*

4. The New Generation

The submarines that made up the anti-carrier programme belonged, in technological terms, to the first generation of modern Soviet submarines. The 'Echo II' was simply a modification of the 'Echo I', and propulsive power was provided by nuclear reactors of the earliest type. The 'Juliett', although purpose-built and not, as stated in some sources, a modification of the 'Foxtrot', was essentially traditional in design as regards hull-form and propulsion. Only in its missile system and the associated electronics did it break new ground.

The conservative technical aspects of both designs are hardly surprising given the dramatic nature of the changes in defence policy of 1959–60, and the haste with which these policy decisions had to be implemented by the Soviet Navy. The inertia built into the Soviet military-industrial system ensured that the new submarines would have to be based on designs already in production to avoid total disruption in the factories and the shipyards.

The anti-carrier programme therefore constituted yet another short-term expedient solution, albeit on a far grander scale than previously. However, the policy decisions reached in 1959–60 also laid the foundations for longer-term developments which would have major implications for the Soviet Navy. In particular, it was around this time that the decision was taken to develop nuclear reactors of improved design which would power a new generation of Soviet submarines. Of even greater importance was the decision to expand future production of nuclear-powered submarines from the current maximum of six to a total of ten per year. This figure was to be attained by providing nuclear construction facilities at a further three shipyards.

One of the three new nuclear building yards was to be the inland Gorky Shipyard, which was currently producing 'Julietts' for the anti-carrier programme. The second was the Admiralty Shipyard at Leningrad, which had not built submarines previously, but which had 'nuclear' experience because of its association with the Soviets' first nuclear-powered ice-breaker, the *Lenin*, which had been completed in September 1959. The third was the Sudomekh Shipyard at Leningrad, which was currently involved in series production of the 'Foxtrot' diesel attack boat. Sudomekh was to continue its traditional association with experimental types, and appears to have been given the brief for the development of unconventional propulsion systems and new construction materials for future generations of nuclear submarine.

With the implementation of this particular policy decision, every major Soviet submarine building yard would be capable of nuclear construction. Given that nuclear-powered submarines of the day were generally credited with a service life of 30 years, the clear implication was that within a similar period of time the Soviets would be able to build and sustain an all-nuclear force of 300 submarines.

It is of interest to note that the Soviets appear to have considered this long-term goal achievable only if their previous policy of concentrating nuclear submarine construction in the Northern and Pacific Fleet areas were compromised. The decision to convert two of the traditional Baltic yards and the inland shipyard at Gorky to nuclear construction almost certainly reflects domestic economic factors, rather than military requirements. In particular it would undoubtedly have been difficult to build up a skilled labour force of sufficient size at the unpopular Arctic and north-west Pacific locations. The Soviet Navy would therefore have to rely on inland waterways to transfer the completed submarines to their primary operational areas.

The debate continues

While these decisions were being taken the debates about strategy continued. Within a year of the Soviet policy changes of 1959–60, further reconsideration was made necessary by policy changes in the United States. In early 1961, in an atmosphere charged with concern regarding an alleged 'missile gap' developing in favour of the Soviet Union, President Kennedy decreed that the production rate of solid-fuel ICBMs should be doubled, and the US Navy's Polaris submarine programme sharply accelerated. The number and variety of nuclear delivery systems becoming available raised the possibility of a 'limited' nuclear war, in which some systems would be withheld as a strategic reserve to provide a bargaining counter.

These ideas were openly discussed by American defence experts, and their arguments had a considerable impact within the Soviet Union, where the influential Marshal Malinovsky, in a speech in October 1961, called for a nuclear warfighting policy.

The balance of the delivery systems in the so-called nuclear 'triad' (ground-launched, air-launched, and sea-launched) on which US strategic military policy was based was very different from that envisaged in the Soviet policy decisions of 1959–60. In contrast to the Soviet decision to downgrade both the Soviet Navy's nuclear role and the importance of the strategic bomber force in favour of ICBMs in fixed land-based silos, submarines would accommodate one-third of all US nuclear delivery systems by 1966, and one half by 1970.

Some Soviet strategists felt that the Soviet Union should reply in kind to these developments. The technology, in the form of the SS-N-5 missile and missiles with longer range under development by 1960, was already there, and it is reported that in 1961 Gorshkov tried to interest Krushchev in new submarine designs armed with 16 or 24 missiles similar to those being built for the US Navy. Krushchev strongly resisted these projects, however, arguing that the Soviet geo-strategic situation was very different from that of the United States, and that different military systems were therefore required.

In particular, Krushchev was concerned about the growth in US anti-submarine capabilities, which would place Soviet ballistic submarines operating off the coasts of the United States at considerable risk. Moreover, whereas the US Polaris boats could operate from Allied bases in Europe such as Holy Loch (UK) and Rotá (Spain), there was no similar possibility of 'forward deployment' for Soviet ballistic missile submarines, which would therefore be able to spend only a fraction of the time on patrol that could be sustained by the Western boats.

The accuracy of the Soviet sea-based ballistic missiles was still a major concern. The Soviet submarines lacked the equivalent of the US Navy's Ship Inertial Navigation System (SINS), and therefore had great difficulty in determining their precise position. During the early 1960s the Soviets therefore embarked on a major ocean survey programme aimed at producing detailed charts of the sea bottom in various key areas. Subsequently Soviet ballistic submarines were able to use these charts, in conjunction with celestial observation, for navigation.

Regular patrols off North America began only in 1966. Because of the known location off both the west and east coasts of large fixed underwater hydrophones belonging to the US SOSUS underwater surveillance system, the submarines were generally placed in holding areas about a day's sailing from the launch zones. The areas west of the Azores and east of Nova Scotia were those favoured in the North Atlantic, and the area west of Hawaii in the Pacific.

The appearance in 1967 of a completely new type of Soviet ballistic missile submarine, which subsequently received the NATO designation 'Yankee', marked a major change in policy direction.

The 'Yankee'

The first unit of the 'Yankee' class was laid down at Severodvinsk probably in the latter half of 1964, but certainly no later than early 1965. The timing is significant, as it was in 1964 that the Soviet Communist Party's Central Committee finally lost patience with the radical policies of Nikita Krushchev, which had wrought havoc on the continuity of both industrial and agricultural production, and dismissed him. Leonid Brezhnev, who took over as Party Secretary, and Alexei Kosygin, who became Prime Minister, were apparently selected by the Central Committee for their pragmatic conservatism, and moved quickly to undo Krushchev's wilder schemes.

Krushchev's approach to military strategy was not dissimilar to that of Winston Churchill. He had a tendency to see everything in black and white, and was prone to making sweeping judgements about issues that were inherently complex. His successors were more cautious and less 'internationalist'. They were more concerned with the defence of the Soviet Union and of its immediate sphere of interest than with conducting a diatribe against Western imperialism on the world stage.

Whereas Krushchev was prepared to conduct a constant radical reassessment of the organization and equipment of the armed forces, the new leadership wanted to consolidate previous technological gains and to stabilize military-industrial production. It is of interest to note that once the new generation of submarines was laid down in the mid-1960s, the same types were to remain in production into the 1980s – a remarkable turn-about considering the instability that characteristized Krushchev's years in office. Only incremental improvements to these designs were permitted, in order to ensure minimal disruption to production goals.

Krushchev had been dismissive of the Soviet Navy's claim to a strategic nuclear role, but Gorshkov's proposals for a force of more advanced ballistic missile submarines now received a more sympathetic hearing. The new leadership was concerned about the increasingly aggressive stance of the United States, and in

Right: *The 'Yankee' was the first Soviet ballistic missile submarine modelled on the US Navy's Polaris boats, with the missile tubes located in the hull, not in the conning tower. Note the prominent unpainted sonar 'window' in this early view. (US Navy, 1971)*

particular about the new theories of limited nuclear war. The threat of 'counter-force' strikes against Soviet land-based missiles made the construction of a 'reserve force' comprising ballistic missile submarines a prudent investment for the future. The SS-N-6 missile which was then under development suffered from the same range deficiencies as its predecessors, and would therefore have to be launched from waters dominated by Western anti-submarine defences. However, it was expected that these defences would be seriously eroded during the earlier stages of a nuclear conflict, prior to the deployment of the Soviet missile submarines to their designated patrol areas.

The 'Yankee'/SS-N-6 combination has been described as an 'interim system', and in developmental terms it most certainly was. Even while the first units were building, work was proceeding on a new missile that would have a maximum range more than three times that of the SS-N-6, and could therefore be launched against the US mainland from within Soviet-dominated waters in the Arctic and the north-west Pacific. The new missile would be significantly larger than the SS-N-6, and could not therefore be accommodated in the launch tubes of the 'Yankee'.

In view of this, the sheer scale of the 'Yankee' programme is somewhat surprising. By 1969, when the 'Yankee' was in full production at both Severodvinsk and Komsomolsk, production was proceeding at the rate of six boats per year, and in 1970 no fewer than eight were completed. Given that the total nuclear building capacity of Soviet shipyards was no more than ten boats per year, this was a major programme, and one which dwarfed the cruise missile and torpedo-armed submarine programmes which were proceeding concurrently.

Moreover, the way in which the design was rushed into production shortly after Krushchev's dismissal from power suggests that the 'Yankee' programme was accorded the highest priority. The original design almost certainly dates from the early 1960s, when models of similar submarines are reported to have been shown to Krushchev by Gorshkov. The propulsion plant is thought to be based on a nuclear reactor developed for the cruise missile submarines of the 'Charlie' class, the first unit of which appeared some months after the first 'Yankee' was observed. In strategic terms the 'Yankee' therefore appears to have meant more to the Soviets than is suggested by the term 'interim system'.

'YANKEE I' CLASS

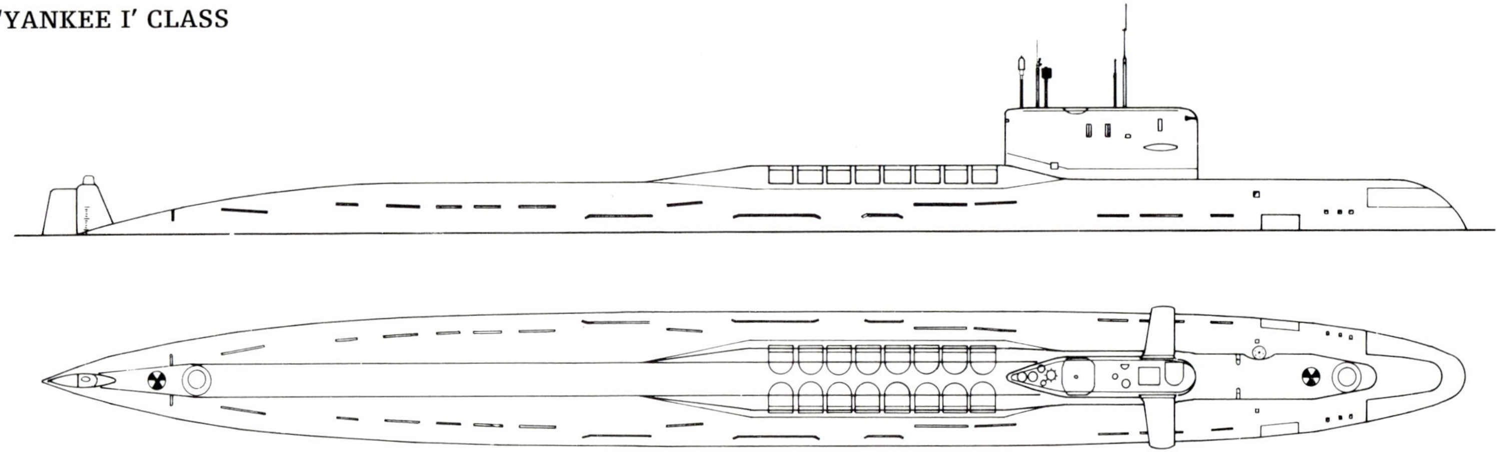

'YANKEE' CLASS

Soviet designation: Podvodnaya Lodka Atomnaya Raketnaya Ballisticheskaya (PLARB) (Ballistic Missile Submarine, Nuclear-Powered)

Construction:
Built at Shipyard No. 402, Severodvinski; Komsomolsk-na-Amur.
34 units completed 1967–74: two in 1967, four in 1968, six in 1969, eight in 1970, six in 1971, five in 1972, two in 1973, one in 1974.

Displacement:
Surfaced: 7,900 tons
Submerged: 9,600 tons

Dimensions:
Length: 130m (430ft)
Beam: 12m (39ft)
Draught: 8.8m (29ft)

Propulsion:
Two pressurized water-cooled reactors driving geared steam turbines; two shafts with 5-bladed propellers; 44,000shp for 27 knots max.

Complement 120

Weapons:
Sixteen SS-N-6 Serb ballistic missiles in vertical launch tubes: six bow tubes of 533mm (21in) diameter for anti-ship/anti-submarine torpedoes; maximum capacity eighteen torpedoes.

Sensors:
LF active/passive sonar; search and attack periscopes; Snoop Tray surface surveillance radar; Brick Pulp/Brick Spit ESM; Park Lamp DF loop; Cod Eye radio sextant.

Communications:
HF and VHF radio.

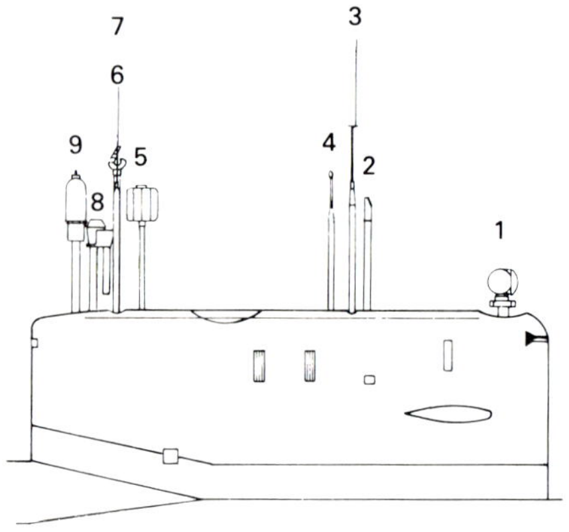

'Yankee'
1. Cod Eye radio sextant
2. search periscope
3. HF communications mast
4. attack periscope
5. Park Lamp VLF reception
6. Snoop Tray surface surveillance radar
7. VHF/UHF communications mast
8. snorkel for emergency diesel
9. Brick Group ECM

The problem of the 'Yankee' mission, which has been the subject of some debate in Western defence circles, is complicated by the development of a Soviet terminally guided, submarine-launched ballistic missile, the SS-NX-13. The SS-NX-13 was apparently an attempt by the Soviet Navy to develop ballistic missiles for the tactical anti-ship role. Although not tested until 1972, the concept may predate the SS-N-16 missile. Indeed, it may have originally been envisaged that it would replace the SS-N-5 missile aboard submarines of the 'Golf' and 'Hotel' classes, thereby replacing the strategic mission of those submarines with a tactical 'anti-carrier' mission. This would seem to accord closely with Krushchev's own ideas about Navy missions as formulated following the 1959–60 defence review.

The SS-NX-13 development programme appears to have been plagued with technical difficulties, and the system was finally abandoned in 1973. Nevertheless, the continuation of the programme after Krushchev's fall from power and into the early 1970s has led some commentators to speculate that it was intended that

Above: *Note the different pattern of free-flood holes on the outer casing, which mark out this 'Yankee' as one of the later units of the class. The fin-mounted diving planes, unique to the 'Yankee' and to their lineal successors of the 'Delta' class, are particularly prominent in this view. (MoD, 1987)*

the 'Yankee' should convert to this missile as later types of SSBN armed with the long-range SS-N-8 became available.

The first 'Yankee' patrols in the North Atlantic began in June 1969. Typically one SSBN was stationed to the north of Bermuda, with a second boat to the south of that island. Older boats of the 'Hotel' and 'Golf' classes were stationed east of Nova Scotia and west of the Azores. Patrols in the Pacific, west of Hawaii, are reported to have commenced in 1970. By the mid-1970s, when the last of the 'Yankees' was completed, the deployment ratio was about three to one in favour of the Northern Fleet. However, as more modern SSBN types were completed a number of 'Yankees' were transferred to the Pacific.

It has been suggested by some commentators that the mission of these submarines was to launch minimum-warning strikes against US strategic bomber bases, firing their missiles from short range using depressed trajectories. However, no Soviet tests involving depressed-trajectory firing have been reported, and it appears more likely that the patrolling submarines had a purely deterrent mission, and that their missiles were targeted primarily on naval installations such as ports and harbours. Targets such as these would make it unnecessary for the submarine to approach close to the US mainland, where the threat from anti-submarine forces would be greatest, and would have been more acceptable in 'political' terms, given that strikes against the major US inland cities and military installations continued to be regarded as the province of the Strategic Rocket Forces.

Technical specifications

The basic layout of the 'Yankee' bears a striking resemblance to that of the US Navy's Polaris boats. The missile compartment, housing sixteen launch tubes, is located amidships, with the fin immediately forward of it. The control centre, torpedo room and accommodation spaces are forward of the missile compartment, and the machinery spaces abaft it. Cruciform after control surfaces were adopted for the first time in a Soviet submarine while, in what constituted a significant break with previous practice, the forward diving planes were located on the fin.

The superficial similarities between the 'Yankee' and its US counterparts are deceptive however. A closer comparison of the two types reveals fundamental differences in submarine design philosophy which can help to explain much of subsequent Soviet SSBN development.

The most useful comparison is perhaps between the 'Yankee' and the US Navy SSBNs of the *Lafayette* class, the last of which was completed in 1967, the year that the first 'Yankee' entered service. Overall length, at 130 metres, is the same for both classes, but there is a 2-metre difference in beam (10.1 metres in the *Lafayette*, 12 metres in the 'Yankee'). This is a direct consequence of the Soviet persistence with double-hull construction and results in a submerged displacement of 9,600 tons, as compared with only 8,250 tons for the American boat. Double-hull construction also brings with it the need for large numbers of free-flood apertures in the hull to ensure that the submarine can dive quickly if

located by a hostile aircraft. Moreover, in accordance with standard Soviet practice, the fin of the 'Yankee' is some 3.5 metres longer than that of the *Lafayette*, and the bow is more fully rounded.

All of these factors combine to create additional resistance to movement through the water, and had the Soviets adopted a propulsion system of similar output to that of the *Lafayette*, the 'Yankee' would have had a maximum sustained underwater speed probably not exceeding 20 knots. For the American boats, a relatively low transit speed would have been more acceptable, because of the proximity of their forward bases to their patrol areas. The Soviet boats, however, would have to make a transit of 2,500–3,000 nautical miles in order to reach their patrol areas, and would have to evade NATO barrier anti-submarine defences *en route*.

The only solution available to the Soviets was to provide a significant increase in horsepower. Just how significant this increase was is the subject of some disagreement, with British sources citing a figure of 35,000shp, while some US sources put the figure as high as 50,000shp. The latter figure seems excessive, given that the *Lafayette*s and their European counterparts are generally credited with an underwater speed of about 24 knots on only 15,000shp. It also conflicts with the suggestion that the 'Yankee' propulsion system was based on the reactor installed in the cruise missile submarines of the 'Charlie' class, which is generally rated at 15,000shp. It is often forgotten that nuclear reactors, like gas turbines and diesel engines, have a fixed maximum output. The figure of 50,000shp would imply that a separate reactor was specifically developed for the 'Yankee' – a suggestion that contradicts the theory that the design for the class was drawn up in a hurry on the basis of earlier preliminary studies.

What can be stated with some degree of certainty is that the propulsion system of the 'Yankee' employs two nuclear reactors, and that these occupy considerably more internal volume than the single reactor of the *Lafayette*. A comparison of the way in which the length of the two vessels is distributed reveals that the distance between the after pair of missile launch tubes and the after edge of the vertical rudder is 54 metres in the *Lafayette*, and 66 metres in the 'Yankee' – an increase of 20 per cent. This results in some compression of the space available for the missile compartment, which is 27m long in the *Lafayette* but only 21m long in the 'Yankee'. Had the SS-N-6 missile been of

Above: *A recent view of an SSBN of the 'Yankee' class. Note the prominent bulged fairing on the side of the conning tower, reminiscent of those fitted in SS-N-12 modifications of the 'Echo II' and 'Juliett'. (TASS, 1987)*

Right: *In 1986 a 'Yankee' on patrol in the North Atlantic was shaken by an explosion in one of its missile tubes. The liquid propellant of the SS-N-6 missile is thought to have been responsible. The submarine later sank. Apart from the open missile tube and the vertical position of the hydroplanes – testimony to the haste with which the submarine was obliged to surface – the photograph shows a number of interesting features. This unit has received not only the bulged fairings noted in the caption to the TASS photograph, but also displays the updated sensor outfit of the 'Yankees' which remain in service, including a modified version of the Pert Spring satellite navigation antenna, distinguished from the standard model by the side plates suspended beneath the upper cone. Also visible at the after end of the conning tower is the snorkel for the auxiliary diesel, the nuclear propulsion plant presumably having been closed down as a consequence of the explosion. (US Navy, 1986)*

comparable diameter to the US Navy's Poseidon, which the launch tubes of the *Lafayette* were designed to accommodate, there would have been sufficient space for only twelve launch tubes in the 'Yankee'.

This provides yet another illustration of differences in design philosophy between the US Navy and the Soviet Navy. Every class of US ballistic missile submarine has been designed with spare tube capacity to accommodate future missiles. Conversely, new missiles under development have frequently been deliberately sized to enable older submarines to convert to them without drastic structural modifications.

Soviet ballistic submarines, on the other hand, are generally designed around a particular missile. Moreover, Soviet missile scientists do not appear to be constrained by the size of the launch tubes aboard existing submarines, and missile dimensions tend to be determined by performance requirements. As in other areas of Soviet manufacturing industry, the producer (in this case the missile design bureau) decides what he is going to produce, and the consumer (the Navy) just has to find some way to accommodate it.

Consequently, the 'Yankee' was unable to convert to the incomparably superior SS-N-8 missile when the latter became available, and the 'Delta I', which was essentially a modification of the basic 'Yankee' design, was able to accommodate only twelve of the new missiles.

Weapons and sensors

The SS-N-6 missile (NATO code-name: Serb) which equipped the 'Yankee' was a logical development of the SS-N-5 Sark. The principal improvements were in range and in payload. Later variants of the SS-N-5 had a maximum range of 900 nautical miles carrying a warhead estimated at 800KT, whereas the figures for the SS-N-6 were 1,300nm and 1MT respectively. The Mod 2 variant of the SS-N-6 missile, which entered service in 1972, would increase range by a further 300nm, while the Mod 3 which entered service shortly afterwards would have a range identical with the Mod 2 and two re-entry vehicles (MRVs).

Estimates of the length of the SS-N-6 missile have varied between 10 and 12 metres (the difference between the height of the respective hull casings of the 'Yankee' and the 'Delta' suggests a length of 11–11.5m); diameter is generally stated to be 1.65 metres. The

missile is therefore both longer and slimmer than the US Navy's Polaris and Poseidon.

The bow section of the 'Yankee' houses six torpedo tubes of standard 533mm diameter, capable of firing the Type 53 pattern-running anti-ship torpedo and a new electric anti-submarine torpedo with acoustic homing (see 'Victor'). The size and configuration of the unpainted sonar 'window' wrapped around the upper section of the bow suggests a sonar outfit broadly similar to that of the 'November'.

The mast-mounted sensors, however, belong to a different generation. The search and attack periscopes, located centrally, appear to be little different from those on earlier boats, and the Snoop Tray surveillance radar and mast-mounted HF and VHF/UHF aerials are also standard. However, the Quad Loop DF aerial of earlier classes was superseded by a new model capable of VLF reception and designated Park Lamp, and the Stop Light ESM antenna was replaced by Brick Pulp/ Brick Spit (also known as Brick Group), a new ESM sensor comprising a tall radome with two prominent rectangular plates suspended beneath it.

Other new sensors were related to the special mission of these submarines. There was a large hatch at the forward end of the conning tower which concealed a Cod Eye celestial sight, and an even larger hatch between the periscopes and the after group of sensor masts, the contents of which have not been revealed.

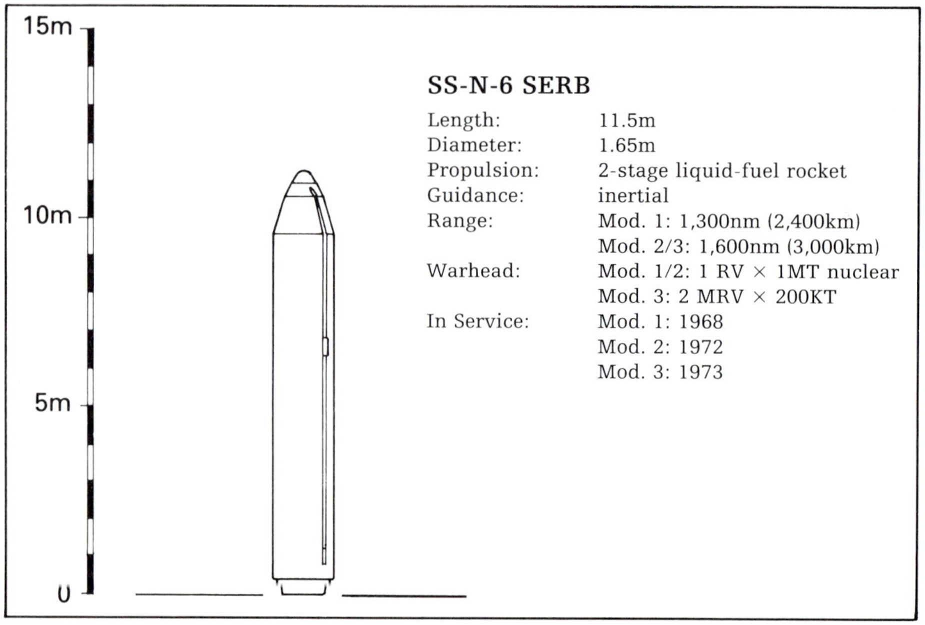

Later variants

During the early 1970s a single unit of the 'Yankee' class was taken in hand for conversion to a new configuration. The missile compartment was modified to accommodate twelve larger missile tubes for a new, developmental ballistic missile which subsequently received the NATO designation SS-N-17 Snipe.

The SS-N-17 was the first Soviet SLBM to employ solid-fuel propulsion, and should therefore be regarded as the precursor of the giant SS-N-20 missile which equips the 'Typhoon'. With a length variously estimated at between 10.6 and 11.1 metres and a diameter of 1.8 metres, the SS-N-17 is credited with a maximum range of 2,000 nautical miles.

Opposite page: *The single 'Yankee II' variant. Note the higher missile compartment and the 'steps' abaft it, which give this submarine an external configuration similar to that of the 'Delta I' (MoD)*

Above: *A recent view of a 'Charlie I' running on the surface. Note the 'collar' added at the base of the conning tower to improve the waterflow. (MoD)*

The appearance of the 'Yankee II' when the submarine emerged from refit in 1976 was strikingly similar to that of the 'Delta I'. The missile casing was higher than that of the original 'Yankee', and descended in graduated steps towards the stern section. New sensors included a LF buoy, for which doors were set into the casing immediately abaft the missile compartment, and a Pert Spring satellite navigation antenna.

The 'Yankee II' was almost certainly intended as the prototype for other 'Yankee' conversions, but, implementation of the SALT I Treaty, which imposed a 950-missile ceiling on the Soviet SSBN force, caused the programme to be abandoned. Indeed, in order to comply with the terms of the treaty, 'Yankees' would have to be taken out of service so that more advanced types of ballistic missile submarine could be completed. The 'Yankee II', although retained in first-line service following test-firings with the SS-N-17 missile, remained a one-off conversion.

From the early 1980s the original 'Yankee I's began to decommission. Sixteen had been taken out of service by the end of 1987. Most remained alongside with their missile tubes disabled. A prototype SSN conversion with a modified bow similar to that of the 'Victor' was not pursued further. A second unit was modified to conduct trials of the SS-NX-24 cruise missile (*see* Chapter 9).

From the mid-1980s a variant known as the 'Yankee Notch' entered service. In this conversion the missile compartment was replaced by a longer section with horizontal launch tubes. The hull has been 'pinched' at the forward end of the new weapons compartment, apparently to accommodate the muzzle ends of the tubes. What is housed within the new weapons section is not yet clear. Suggestions include the SS-N-21 cruise missile, with a possible complement of between 20 and 40 missiles, or mines for the defence of the Soviet SSBN bastion in the Arctic. Unlike the other two conversions outlined above, it is reported that other redundant 'Yankee' hulls are being converted to this configuration.

The 'Charlie'

The second major submarine type to enter service in 1967 was the 'Charlie', a nuclear-powered cruise missile submarine armed with the revolutionary SS-N-7 'pop-up' missile.

Earlier Soviet cruise missile submarines equipped with the long range SS-N-3 Shaddock had to launch their missiles from the surface, and were therefore vulnerable to pre-emptive strike by carrier-borne attack aircraft. Their torpedo-armed counterparts of the 'November' and 'Foxtrot' classes could close the task force while running submerged, but their torpedoes had to be fired at relatively short ranges (typically 5,000–15,000 metres) in order to have some hope of success. They would therefore have to run the gauntlet of increasingly effective anti-submarine defences, which would shortly include helicopters operated by the surface escorts. Even if nuclear-tipped torpedoes were employed, there was considerable risk of damage to the submarine firing the torpedoes.

With the advent of the SS-N-7 missile (NATO code-name: Starbright), which could be fired from underwater at a maximum range estimated at 30–35 nautical miles by Western sources, the Soviet Navy was able to combine the advantages of underwater approach with a stand-off weapon which could be fired from well outside the protective ring of carrier escorts. The proximity of the launching submarine to its target also ensured greater accuracy, because of the much-reduced flight time. This eliminated the need for mid-course guidance, which was undoubtedly the weak link in the SS-N-3 targeting system, and made the SS-N-7 missile a more suitable vehicle for a conventional high-explosive (HE) warhead.

Some sources credit the SS-N-7 with an alternative 200KT nuclear warhead, but there is by no means universal agreement on this point. Soviet strategic thinking in the post-Krushchev era was already moving towards the concept of limited war, possibly beginning with a protracted conventional phase. The American super-carriers would undoubtedly be the major implements for the projection of US power during such a phase, and would therefore have to be countered effectively with sub-nuclear weapons.

Technical characteristics

In tactical terms the 'Charlie' is derived from the SSNs of the 'November' class rather than the 'Echo' SSGNs. Soviet sources have spoken of the SS-N-17 missile as a 'flying torpedo', thereby adding weight to this proposed line of succession. The 'Echo' was intended to operate on the surface, and was therefore given a traditional hull-form designed for good sea-keeping and stability. The hull-form of the 'Charlie', like that of the 'November', was designed for underwater performance.

The major difference between the 'Charlie' and the 'November' was in their respective lengths, the 'Charlie' being some 16 metres shorter than its predecessor. This has led to suggestions that the dimensions of the 'Charlie' were deliberately restricted in order to allow construction at the Gorky Shipyard,[1] whence the submarines were to be transported to the Northern Fleet area via inland waterways. Overall dimensions were undoubtedly a concern, as Gorky had previously built only diesel-powered submarines of the 'Whiskey' and 'Juliett' classes (the length of the latter was 90m), but it should be noted that the later 'Charlie II' variant was to have a length of 102m, and that the experimental 'Papa', completed in 1970 at the same shipyard, is of similar length to the 'November'.

More influential in the move to smaller dimensions was the reduction from two nuclear reactors in the 'November' to a single reactor in the 'Charlie'. The reactor was of new design, and was probably more compact than the first-generation 'HEN' reactor in-

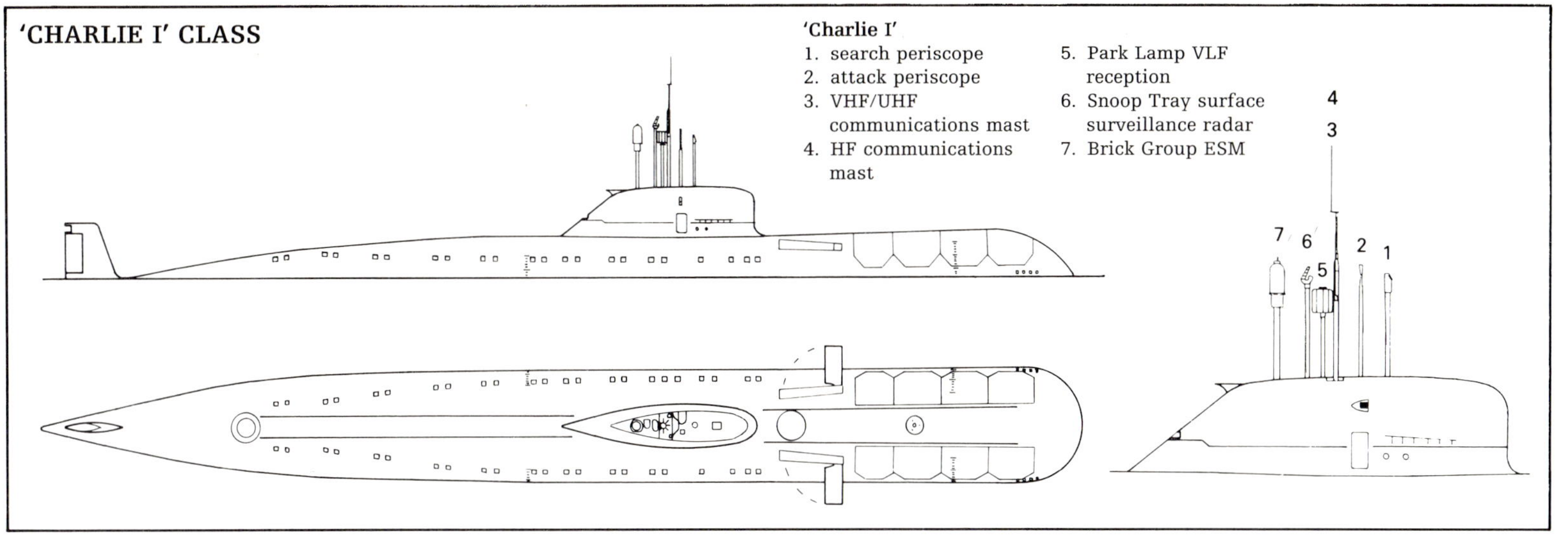

'CHARLIE I' CLASS

Soviet designation: Podvodnaya Lodka Atomnaya Raketnaya Krylataya (PLARK) (Cruise Missile Submarine, Nuclear-Powered)

Construction:
Built at Krasnaya Sormovo Shipyard, Gorky.
Twelve units completed 1967–73: one in 1967, two in 1968, one in 1969, three in 1970, two in 1972, three in 1973.

Displacement:
Surfaced: 4,000 tons
Submerged: 4,800 tons

Dimensions:
Length: 94m (308ft)
Beam: 10m (33ft)
Draught: 8m (26ft)

Propulsion:
One pressurized water-cooled reactor driving geared steam turbines; one shaft with 5-bladed propeller; 22,000shp for 24 knots max.

Complement 80

Weapons:
Eight SS-N-7 Starbright anti-ship missiles in fixed angle launch tubes; six bow tubes of 533mm diameter for anti-ship/anti-submarine torpedoes; maximum capacity fourteen torpedoes.

Sensors:
LF active/passive sonar; search and attack periscopes; Snoop Tray surface surveillance radar; Brick Pulp/Brick Spit ESM; Park Lamp DF loop.

Communications:
HF and VHF radio.

Opposite page: *Of the three types which made up the 'second generation' of Soviet nuclear-powered submarines, it was the 'Charlie', with its SS-N-7 'pop-up' missiles, that created the greatest initial impact. Note the exceptionally full bow section, in which the angled launch tubes were located. (MoD, 1975)*

stalled in the submarines of the 'Hotel', 'Echo' and 'November' classes.

A major innovatory feature, at least as regards Soviet submarine design, was the adoption of a single centre-line shaft on the pattern of contemporary Western nuclear-powered submarines, with the propeller abaft the after control surfaces. Both the 'November' and the 'Yankee', which had propulsion systems powered by two reactors, were driven by twin shafts. The adoption of a single shaft was more economical in terms of both space and weight, and made for much quieter operation because of the marked reduction in water turbulence. The turbine machinery and pumps were, like the reactor itself, of more modern design, and were therefore inherently quieter and more reliable. However, the primary purpose of flexible mountings continued to be the reduction of shock damage in the event of an underwater explosion close to the submarine. There appears to have been little conscious effort to reduce machinery noise transmitted through the hull of the submarine.

There was probably inadequate internal volume to permit any form of rafting in Soviet submarines of this generation. Although the external dimensions of the 'Charlie' are similar to those of contemporary Western SSNs, Soviet persistence with double-hull construction meant that the diameter of the inner pressure hull was significantly reduced. Soviet nuclear boats of the period were therefore undoubtedly more cramped than their Western counterparts.

The single reactor of the 'Charlie' is rated at 11.5MW (equivalent to 15,000shp) by the major Western reference sources, which credit these submarines with a maximum speed of 24 knots. This constitutes a reduction of 50 per cent as compared with the installed horsepower of the 'November'. The 'Charlie' was initially credited with a much larger reactor producing 30,000shp for a maximum speed of 30 knots, but this figure was revised downwards during the early 1980s when it was found that these submarines were significantly slower than their torpedo-armed counterparts of the 'Victor' class.

The figure of 15,000shp is based on the assumption that the 'Charlie' and the 'Victor' share a common reactor, with two units being installed in the 'Victor' and only a single reactor in the 'Charlie'. If this theory is correct, the marked similarities between the respec-

tive after sections of these two classes are difficult to explain. Moreover, this theory takes no account of the relationship between the reactor installed in the two classes of attack submarine and that of the 'Yankee', which is credited by all sources with a much higher power rating.

An alternative explanation which has not received sufficient consideration is that the 'Charlie' has a single reactor with a higher power output than the unit that forms the basis of the 'Victor' propulsion plant.

It is of interest to note that the nuclear icebreaker *Lenin* suffered a serious accident involving her original propulsion plant during 1966–7 which resulted in her withdrawal from service. She was subsequently rebuilt at Severodvinsk, recommissioning in 1972. During this rebuild she received a new propulsion plant based on two (*vice* the original three) nuclear reactors, each of which had a power rating of approximately 16MW (equivalent to 22,000shp). Since these reactors appear to have been readily available for installation, it is conceivable that they are derived from the reactor fitted in the 'Charlie'. The 'Victor', on the other hand, may have received a reactor derived from that of the submarines of the 'HEN' generation.

One major attraction of this theory is that it would explain the high horsepower figures with which the 'Yankee' and its successors of the 'Delta' class are generally credited. Doubling up the larger 'Charlie' reactor would produce about 44,000shp, a figure sufficiently high to cater for the power requirements of even the later 'Delta' variants, which would feature significant increases in displacement.

Even 22,000shp constitutes a marked reduction over the twin-reactor plant of the 'Victor'. Presumably high speed was less of a priority for a submarine armed with stand-off missiles, as rapid manoeuvres both before and after an attack would no longer be necessary.

Other features of the design

The eight launch tubes for the SS-N-7 missile are housed within the bow section. Four tubes are disposed on each side of the pressure hull, and are probably installed at an angle of about 60 degrees. The tubes are covered by large square hatches, which are hinged on the inboard side and open upwards to permit firing. When closed, the hatches fit flush with the outer hull casing, thereby eliminating the waterflow problems noted in connection with the SS-N-3 missile installations aboard the 'Echo' and 'Juliett'.

The bow section, however, is exceptionally broad, which must result in a significant increase in drag when the submarines are operating at high speed. The installation of the missile tubes on each side of the pressure hull must also mean that the diameter of the latter is significantly reduced at this point. The 'Charlie' is credited with six bow torpedo tubes of standard 533mm diameter, but the reduction in the cross-section of the pressure hull immediately abaft the tubes undoubtedly imposes limitations on the number of reloads. One reload per tube would seem to be the maximum that could be accommodated in the torpedo handling room.

The fin is low and streamlined, but its vertical forward edge appears to have caused hydrodynamic problems, as a prominent 'collar' has been fitted at the base of the fin in a number of boats since completion. The rationale for the vertical forward edge is not clear, as later units of the 'November' class and contemporary SSNs of the 'Victor' class have a fully streamlined fin with an angled front edge. It is therefore possible that the forward section of the 'Charlie' fin houses some sort of passive array derived from that of the early 'Novembers', which had a fin of similar configuration.

The sensor masts of the 'Charlie' belong to the same

Above and opposite page: *Two starboard-side views of 'Charlie'-class SSGNs running on the surface. The four missile hatches are clearly visible close to the bow, with the angled hatch covering the rectractable diving planes immediately abaft them. Note the attention paid to streamlining, with even the free-flood holes being fitted with closing hatches to ensure a smooth waterflow over the hull when submerged. (US Navy, 1974/ MoD, 1974)*

generation as those of the 'Yankee', although the cruise missile submarines lack the elaborate position-fixing equipment of the latter. The search and attack periscopes are complemented by HF and VHF/UHF communications masts, a Snoop Tray surface surveillance radar, Park Lamp VLF reception and Brick Group ESM. In order to improve waterflow, the apertures for the sensor masts are closed by hinged rectangular hatches – the 'Charlie' was the first class to incorporate this feature, which has been repeated in subsequent Soviet submarine designs.

Western-style cruciform tail surfaces were the natural complement to the adoption of a single propeller, but the low-profile fin was unable to accommodate the forward diving planes, which were therefore located in the conventional way on the forward section of the hull, between the missile tubes and the fin. In accordance with previous Soviet practice the planes were retractable, and when stowed were covered by hinged plates.

Construction

Submarines of the 'Charlie' class were built exclusively at the Gorky yard, where they replaced the diesel-powered 'Juliett' on the construction ways. Production was at a steady rate of two boats per year throughout the late 1960s and early 1970s, with twelve units completed by 1973.

When completed virtually all units deployed initially to the Northern Fleet, although four were later transferred to the Pacific. One of the Pacific units sank off the Kamchatka Peninsula in June 1983, and although subsequently salvaged has not re-entered service.

The appearance of the 'Charlie' had a more dramatic effect on the tactical thinking of the US Navy than any previous development in the Soviet submarine fleet. The limited effective radius of traditional task force ASW operations against submarines armed with stand-off missiles compelled the US Navy to look beyond a circular screen of escorts to other methods of locating and sinking hostile SSGNs before they were able to reach a firing position. A variety of measures were adopted, including the operation of manned helicopters from escorts, the inclusion of a squadron of fixed-wing anti-submarine aircraft in carrier air wings (much to the dismay of the aviation community, as the S-3 Vikings displaced a squadron of attack aircraft), and 'direct support' of the carrier battle group by friendly submarines.

'Direct Support' involved sending SSNs, generally one or two per battle group, ahead of the task force in order to 'sanitize' the waters through which it was due to pass. The concept was difficult to realize because of the high speed at which the task force would be expected to operate in hostile waters. The supporting submarines would also need to operate at high speed in order to stay ahead of the task force. High speed for the nuclear attack boats implied noisy operation, however, and would make it difficult for them to use their delicate passive sensors effectively. The theoretical solution was a 'sprint and drift' mode of employment, but this required a higher margin of speed superiority over the carrier battle group. The US Navy was therefore compelled to build a new type of submarine, the *Los Angeles*, capable both of very high underwater speed and of more silent high-speed operation than its predecessors.

The other problem inherent in the concept of 'direct support' was that of communications, as the implication was that the submarine would have to stay in close contact with the carrier battle group. The only effective way of coping with this problem was for the battle group to follow a predetermined course, and for the

submarine to come close to the surface at regular intervals to enable it to receive any course updates and instructions.

The US Navy also had to cater for the possibility that despite all these measures, the 'Charlie' might well remain undetected and succeed in launching its missiles. This led to major programmes to develop effective electronic countermeasures to jam or decoy the missile homing radar.

With the benefit of hindsight, it is possible to state that the effect of the 'Charlie' on the US Navy was out of all proportion to the number of submarines involved. Contrary to Western expectations, the 'Charlie' construction programme continued to be a relatively low-key effort, and after the early 1970s production declined to only a single boat per year. It finally petered out altogether, as we shall see in a later chapter.

Moreover, in terms of missile accuracy, the 'Charlie' remained a relatively blunt-edged weapon. It is no easy matter to determine the precise location, composition, and formation of a task force using only passive underwater sensors. The 'Charlie' would therefore have been just as dependent as its predecessors on the Soviet ocean surveillance network for targeting information. Detailed information regarding enemy forces would have to be communicated to the submarine via high-frequency radio. The submarine would have to operate close to the surface and raise a communications mast in order to receive these transmissions, and would then be vulnerable to detection by anti-submarine air patrols flown from the carrier. Two-way communication to request further updates would make the submarine's own transmissions liable to interception.

In the absence of precise details of the task force formation, the 'Charlie' would have to launch its missiles in the general direction of the enemy and hope for the best. The missiles would tend to lock on to the first surface unit encountered, and this would most likely be a 'low-value' escort in the outer screen. A solution to the problem of over-the-horizon targeting was therefore still some way off.

Above: *Composite overhead view of a 'Charlie I'. Note the broadening of the bow section and the relatively small, streamlined fin. Only the compass platform is open, all apertures for sensor masts being closed by hinged hatches. (US Navy, 1974)*

The 'Victor'

In contrast to the attention accorded to the 'Yankee' and the 'Charlie' in the West, the appearance of a third new type of nuclear-powered submarine, the 'Victor', attracted little comment. It was initially assumed that this torpedo-armed attack boat was the natural successor to the 'November', with a general-purpose anti-ship mission.

It was to be many years before Western commentators came to realize that the 'Victor' was a fundamentally new type, designed for a mission that was alien to previous Western assumptions regarding the purpose of the Soviet submarine force. The importance of the 'Victor' to the Soviet Navy became apparent only during the late 1970s, when production of the later 'Victor III' variant actually increased as SSBN construction declined. Moreover, of the three Soviet nuclear submarine types that entered service in 1967–8, only the 'Victor' spawned true linear successors in

the new generation of submarines that appeared during the 1980s. In contrast, production of the 'Charlie' peaked during the early 1970s, and then tailed off dramatically; in tactical terms, the 'Charlie' constituted something of a 'dead end', and no natural successor appeared once construction was terminated in the early 1980s.

Western failure to make an accurate assessment of the mission of the 'Victor' can be attributed to the same frame of mind responsible for the erroneous assessment of the SS-N-14 missile carried by the cruisers of the 'Kresta II' class as an anti-ship weapon (the SS-N-10). Western Intelligence in the late 1960s was convinced that the rationale for Soviet naval expansion was to wrest control of the seas from NATO. It therefore naturally followed that the primary mission of both the 'Kresta II' and the 'Victor' was anti-surface warfare. It was not until the mid-1970s that the West began to recognize that the Soviets also had legitimate defensive concerns.

In external appearance the 'Victor' was unremarkable. Of almost identical dimensions to the 'Charlie' SSGN, the attack submarine clearly belonged to the same generation. The broad 'tear-drop' hull, with its low streamlined fin, its single shaft allied to cruciform tail surfaces, and its retractable forward diving planes located immediately forward of the fin, showed the same design processes at work. To all intents and purposes, the submarine appeared to be a 'Charlie' without missiles.

Appearances are often deceptive, and this was particularly true of the 'Victor'. Certainly there were no sinister hatches concealing missile launch tubes, but there were other features, some of which were not immediately apparent, that marked out the 'Victor' as a much more capable submarine than its predecessors.

In addition to the single 5-bladed propeller which drives the 'Victor' at higher speeds, there are two small 2-bladed propellers located on the after horizontal control surfaces. These are driven by electric motors and are used for quiet operations at creep speed.

The outer hull casing has a rubberized anechoic coating designed to absorb hostile sonar transmissions. Close-up photographs have revealed that the bow section is covered by a rough, textured coating

Right: *A 'Victor I' running on the surface, showing a good part of her complement of sensor masts. From left to right: search periscope, Brick Group ESM antenna, Snoop Tray surface surveillance radar, Park Lamp DF loop, and HF communications mast. (MoD, 1979)*

'VICTOR I' CLASS

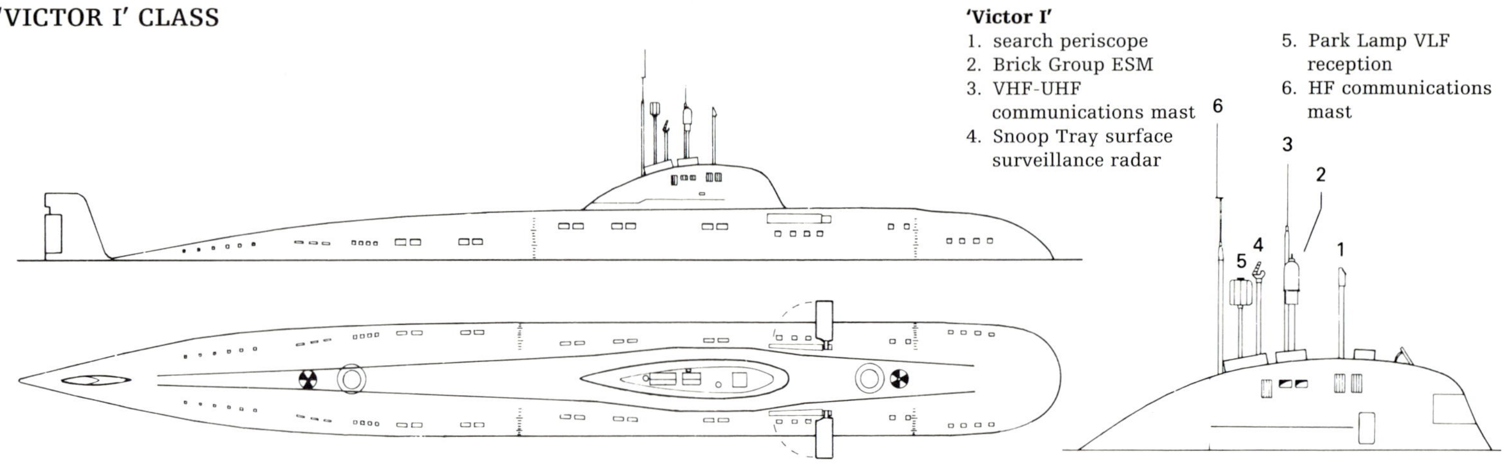

'VICTOR I' CLASS

Soviet designation: Podvodnaya Lodka Atomnaya (PLA) (Submarine, Nuclear-Powered)

Construction:
Built at Admiralty Shipyard, Leningrad. Sixteen units completed 1968–75; two per year.

Displacement:
Surfaced: 4,300 tons
Submerged: 5,100 tons

Dimensions:
Length: 95m (312ft)
Beam: 10m (33ft)
Draught: 7m (23ft)

Propulsion:
Two pressurized water-cooled reactors driving geared steam turbines; one shaft with 5-bladed propeller; two small 2-bladed propellers on stern planes for slow-speed operations; 30,000shp for 30 knots max.

Complement 80

Weapons:
Six bow tubes of 533mm (21in) diameter for anti-ship/anti-submarine torpedoes and SS-N-15 Starfish ASW missiles; maximum capacity twenty-four torpedoes/missiles.

Sensors:
LF active/passive sonar; search and attack periscopes; Snoop Tray surface surveillance radar; Brick Pulp/Brick Spit ESM; Park Lamp DF loop.

Communications:
HF and VHF radio.

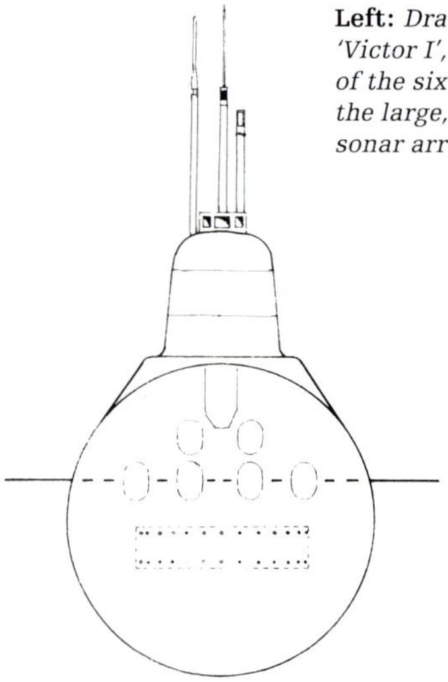

Left: *Drawing of the bow of a 'Victor I', showing the layout of the six torpedo tubes and the large, low-frequency sonar array.*

which appears to have been laid in solid sheets, while from the forward 'hump' in the casing to the stern there are individual tiles, which have received the designation 'Clusterguard' in some sources.

The German Navy experimented with synthetic rubber coatings 4mm thick during the Second World War, and it is certain that samples of this Alberich coating fell into Soviet hands in 1945. However, testing and evaluation of similar anechoic materials for installation on Soviet submarines were apparently not undertaken until the late 1950s and early 1960s, and the submarines of the 'Victor' class may have been the first Soviet submarines to benefit from these developments. Early units of the 'Charlie I' and 'Yankee' classes do not appear to have had this rubberized anechoic coating from the outset (early photographs of the 'Yankee' feature the distinctive unpainted metallic sonar 'window' common in earlier types), although most would receive it during subsequent refits.

The extent to which the rubberized coating is effective in deadening sonar echoes is not clear, but it has been estimated by one US publication[2] that Clusterguard tiles would reduce the acquisition range of a Mk 46 anti-submarine homing torpedo by about 50 per cent. It is also thought that the rubberized coatings improve underwater speed performance, and they undoubtedly deaden machinery noise emissions.

The pattern of free-flood holes in the 'Victor' is very different from that of the 'Charlie'. The major apertures amidships and abaft the fin are larger and of rect-

Right: *Two views of a 'Victor I'. The tear-drop hull-form and low, streamlined fin are superficially reminiscent of the 'Charlie', but there are important differences. The bow section, which houses only a sonar and the torpedo tubes, is not as full, and the face of the conning tower is angled, not straight. Note the distinctive Brick Pulp ESM antenna, just visible in the centre of the conning tower, and the hinged hatches covering the mast-mounted sensors and the forward hydroplanes. (MoD, 1979)*

angular configuration, and appear to be provided with sliding hatch covers which close when the submarine submerges, thereby reducing drag and noise when the submarine is operating at high speed. Hinged hatch covers also close the apertures for the retractable masts atop the fin, as on the 'Charlie'.

The other major new feature of these boats was their large bow sonar array. Earlier Soviet submarines had been fitted with smaller arrays which operated at medium frequency, and therefore could acquire targets at only limited range. These arrays had been mounted within the upper section of the bow, above the torpedo tubes. In the 'Hotel', the 'Golf' and the 'Foxtrot', which were fitted with traditional 'knife' bows for superior performance on the surface, the arrays had to be accommodated in specially configured bulges above the waterline. In submarines with a broader hull-form, such as the 'November', the 'Echo II' and the 'Juliett', the array could be accommodated within the upper bow without any form of bulging, and the same arrangement was adopted for the 'Yankee'.

In the 'Victor', however, the array is much larger than those installed in earlier types, and is located just below the mid-point of the hull, where the bow is at its broadest. The six torpedo tubes are disposed in two rows above the array, the upper row comprising two 533mm tubes and the lower row four tubes (*see* drawing). This unusual arrangement was first revealed to the Western defence press in 1984, when close-up photographs of the damaged bow of a 'Victor I' were taken at Hammamet, Tunisia, following a serious collision between the submarine and a Soviet tanker in the Straits of Gibraltar.

The array comprises a 'crown' of hydrophones, probably in triple staves, some 6 metres in diameter. It is significantly larger than earlier Soviet sonar arrays, and operates at low frequency. If matched with signal processing technology comparable to that available to US Navy submarines of the period (a hypothesis that is nevertheless highly improbable), it would be capable of detecting and tracking targets out to considerable distances. Even if this were not the case, the new array provides an important indication of the direction in which Soviet underwater technology was headed, and

Above: *This close-up of a 'Victor I' shows clearly the rubberized tiles and sheets which make up the hull covering of the submarine. Part of the rubber sheeting wrapped around the face of the fin has been torn, revealing the steel plating beneath. (TASS, 1983)*

Above: *The damaged bow of the 'Victor' which collided with a tanker in the Straits of Gibraltar in 1984. The large 'crown' of hydrophones which makes up the low-frequency sonar transducer can be clearly made out within the torn plating, as can the damaged 'caps' of the six torpedo tubes in the upper part of the bow. (US Navy)*

of the considerable progress already made in developing effective long-range sensors.

A new mission

Clearly, many of the new features first observed in the 'Victor' would enhance the underwater performance of any submarine. The sustained high speeds made possible by a combination of high-performance propulsion machinery and streamlining would be valuable for intercepting hostile surface forces, while anechoic tiling would make the submarine less detectable to surface ship sonars used in the active mode, and would enable the 'Victor' to use its speed and manoeuvrability to avoid homing torpedoes.

Other features, however, notably the auxiliary 'creep-speed' propulsion system and the large low-frequency bow sonar, suggest that the 'Victor' has a primary mission fundamentally different from that of earlier Soviet attack boats – that of hunting and destroying enemy submarines.

The first unit of the 'Victor' class entered service in 1968, some twelve months after the completion of the helicopter cruiser *Moskva*. The latter vessel is generally considered to be the first fruit of a major Soviet Navy anti-submarine warfare programme instituted in the early 1960s in response to the US Navy's deployment of Polaris. From the late 1960s until 1980, a period which saw the Soviet Navy emerge from being a coastal defence force to a major blue-water navy, virtually every major Soviet surface vessel completed had a primary ASW mission. In what was effectively a parallel shift in orientation, Soviet Naval Aviation, which from its inception had been primarily a force of maritime bombers, was to deploy first the Il-38 May, then the Tu-142 Bear F anti-submarine patrol aircraft.

It would have been surprising, given the Soviet predilection for a co-operative Combined Arms solution to any given military problem, had the Soviet Navy not attempted to develop an underwater counterpart to these surface and aerial anti-submarine forces. The outcome appears to be the 'Victor', together with the initiation of a number of programmes aimed at developing more effective underwater sensors and fire

Above: *The layout of the torpedo tubes and the torpedo reloading hatch is particularly clear in this bow quarter view of a 'Victor I'. Note the distinctive 'hump' abaft the bow. (MoD, 1986)*

control systems, and more effective anti-submarine weapons capable of underwater launch.

The 'Victor' combined the first of the new high-performance passive sonars with a new electrically driven homing torpedo, which is reported to have received the NATO designation SET-65.[3] Fired from a standard 533mm torpedo tube, this heavyweight torpedo has a length estimated at 7.8 metres, and has two possible speed settings. At its top speed of 35 knots it has a maximum range of about 10,000 metres, while at 24 knots this range is doubled. Power for the electric motor is provided by silver-zinc batteries, and the torpedo is free-running with active/passive sonar homing and impact/proximity fuzing.

Anti-submarine or pro-submarine?

Moskva and the Large Anti-submarine Ships of the late 1960s were initially intended to hunt down Polaris missile boats in the eastern Mediterranean and the North Atlantic. They were to have been assisted in this task by long-range maritime patrol aircraft, and presumably also by the 'Victor'-class submarine.

Co-operation between aircraft, surface units and submarines is well-established in Soviet tactical doctrine, and there can be little doubt that exercises involving Soviet anti-submarine hunting groups have included submarines operating in direct support. The communication difficulties inherent in such operations have been noted in connection with the US Navy's attempts to provide submarine 'direct support' for its carrier battle groups. Nevertheless, the attempt to harness the potential of the submarine to surface operations has a long (albeit somewhat chequered) history, beginning with the British steam-powered submarines of the 'K' class completed towards the end of the First World War.

It is of interest to note that even after its disastrous experience with the 'K'-boats, the Royal Navy continued to believe that the essential problem was that of relative speed, and with the advent of nuclear power the concept of joint operations was revived. The first nuclear-powered submarines built for the Royal Navy in the 1960s were initially designated Fleet Submarines, and during that period there were few overseas force deployments that did not have an SSN operating in direct support.

The Soviet Navy was therefore by no means alone in believing that the most effective way of hunting hostile submarines was by a combination of forces above, on and beneath the water using a variety of sensors and capable of delivering a variety of weapons. Moreover, Soviet designers had traditionally paid more attention than their Western counterparts to the provision of short-range (VHF/UHF) radio communications and underwater telephones aboard submarines to enable them to co-operate more easily with surface ships and aircraft, and Soviet submariners seem to have had fewer qualms about operating at periscope depth in order to facilitate regular communication with friendly forces.

There is also the possibility that the Soviets envisaged deploying submarines of the 'Victor' class close to the forward bases of the American SSBNs on the Clyde and at Rotá (Spain), where they would pick up the missile boats as they left harbour and shadow them during their patrols. NATO continues to regard any information about Soviet submarine operations as sensitive, and there are therefore no firm indications as to whether 'trailing' has been attempted. All we have to go on are categorical assertions from both the US Navy and from the Royal Navy that not one of their SSBNs has ever been detected while on patrol. If these assertions are true, it follows that if the Soviets have indeed attempted shadowing operations, they have met with little success.

The other mission that suggests itself is a 'pro-submarine' mission in support of the Soviet Navy's own missile submarines. Given the importance of the 'Yankee' SSBN programme to the Soviet Navy, it seems logical to assume that some supportive measures would be necessary to ensure that these submarines arrived safely at their patrol stations off the east and west coasts of the United States. Admiral Gorshkov has stated as much in his book *The Seapower of the State*, where he notes the failure of the German Navy to support its own submarines during their transit through hostile waters to their patrol areas in the North Atlantic.

During the late 1960s and early 1970s the NATO anti-submarine barrier in the so-called Greenland/Iceland/United Kingdom (GIUK) Gap presented the greatest threat to any Soviet submarine attempting to break out into the North Atlantic. Moreover, the aerial domination of the US Navy's attack carriers would have made it extremely difficult to support the deployment of Soviet submarines through the GIUK Gap with surface forces. It is therefore possible that 'Yankee' SSBNs attempting a transit would have had an SSN of the 'Victor' class 'riding shotgun'.

The deployment pattern of the 'Victor' appears to support both the anti-Polaris and the pro-'Yankee' missions. Of the first twelve boats completed all deployed to the Northern Fleet area, and it was only in the mid-1970s that the first units were transferred to the Pacific. The latter move may well have been influenced by the US Navy decision to redeploy some of its older SSBNs of the *George Washington* and *Ethan Allen* classes to forward bases in the Pacific.

Below: *The first unit of the 'Victor' class to deploy to the Pacific Fleet, photographed in the South China Sea in 1974. Note the large, rectangular free-flood holes in the outer casing. (US Navy)*

5. 'Alfa' and 'Papa'

The haste with which the Soviet Navy embarked on the construction of nuclear-powered submarines has already been noted. The preoccupation with other forms of closed-cycle propulsion during the early post-war years meant that the Soviet Navy would be involved in a technological 'tail chase' for many years. Series production of nuclear-powered submarines had to begin without the benefit of developmental prototypes as a matter of military necessity. In consequence, the early Soviet nuclear boats of the 'November', 'Hotel' and 'Echo' classes were a full generation behind the US Navy submarines under construction in the early 1960s.

However, the Soviet submarines of the 'HEN' generation, important as they would be to Soviet force structures during the 1960s and the early 1970s, constituted only the short-term 'contingency' solution to the problem of developing effective underwater weapons systems capable of contesting Western naval supremacy. Smaller, but no less significant, research and development programmes were initiated at the same time to deliver long-term solutions which would use advanced technology to 'leap-frog' over US developments in the field.

Soviet estimates of the future capabilities of nuclear-powered submarines may have been influenced by officially inspired US statements to the effect that within the decade submarines would attain underwater speeds of 45 knots and operate at depths of 3,000 feet. Ironically, subsequent US analysis of future missions and of the force structures needed to perform them was to lead to the abandonment of the fast submarine concept by the US Navy.

It was not that the Americans lacked confidence in their scientific prowess. It was simply that the requisite investment in propulsion technology to develop a fast submarine could not be justified in terms of its utility. The small size of the Soviet surface fleet meant that the requirement for a high-perfomance attack submarine was strictly limited, and funds for such a programme could be procured only at the expense of anti-submarine warfare capabilities, which were judged to be paramount. The US Navy therefore opted to build a series of general-purpose attack boats with an anti-submarine mission bias. In anti-submarine warfare stealth was deemed more important than high performance. Quieting measures therefore took precedence over high speed. Rafting and other forms of flexible mounting for machinery required a larger submarine hull, so that the maximum speed of US nuclear submarine types actually declined between the *Skipjack* (SSN-585), the *Thresher* (SSN-593) and the *Sturgeon* (SSN-637) classes.

The trend towards larger, quieter, slower attack submarines gained even greater impetus during the late 1960s. Even the 'one-off' developmental types of nuclear submarine built for the US Navy during this period, the *Narwhal* (SSN-671, completed 1969) and the *Glenard P. Lipscomb* (SSN-685, completed 1974), were designed as test-beds for quiet, not high-performance propulsion systems. The former was given a reactor employing natural circulation in order to eliminate the noise made by the pumps which circulated the coolant water; the latter tested a nuclear propulsion system which employed turbo-electric drive in order to eliminate the noise generated by steam turbine gearing. Both submarines suffered a significant reduction in speed as compared with the mainstream nuclear boats.

An 'interceptor' submarine

Early US interest in nuclear propulsion for submarines focused on the endurance aspect, rather than on high underwater speed. In effect the US geo-strategic position, separated by two broad oceans from its major allies in Europe and the Far East, demanded a high *transit* speed in order to sustain long distance operational deployments. The endurance required by lengthy patrols could be provided only by nuclear power, and with the development of nuclear reactors suitable for submarine installation, construction of conventional diesel-powered submarines ceased altogether.

The Soviet Navy was equally impressed by this aspect of nuclear submarine propulsion, hence its adoption for the submarines of the 'HEN' generation, which had a primarily strategic mission demanding long range and good endurance.

However, there was another aspect of nuclear propulsion which appealed to the Soviets. The ability of nuclear-powered submarines to sustain very high underwater speed in all weathers over relatively short periods made them potentially ideal 'interceptors' of hostile surface forces. Such a concept accorded closely with traditional Soviet tactical thinking. This was indeed what the Soviet Navy had tried to achieve in the years following the Second World War using German closed-cycle propulsion technology.

The 'interceptor' concept did not require a submarine with long endurance. It could therefore be considerably smaller than the nuclear-powered submarines then under construction. Repairs and maintenance could be carried out almost exclusively at shore bases; a relatively small complement would therefore be required, with consequent economies in the provision of accommodation and dry stores. Sonar capabilities would need to be adequate only to re-acquire a target initially located by aerial reconnaissance or by other elements in the Soviet ocean surveillance network.

The brief for this long-term programme, which would involve the development of much new technology, was accorded to the Sudomekh Shipyard at Leningrad, which had built the first 'Zulu' and was now engaged in production of its successor, the 'Foxtrot'. Sudomekh was not involved in any of the other 'mainstream' nuclear submarine programmes, and would therefore be free from the pressures of production quotas which restricted research and development at other shipyards. Moreover, the design bureau at Sudomekh had been closely associated with innovative propulsion developments since the late 1940s, when the yard had been rebuilt using industrial machinery transferred from Germany.

It is not known precisely when the interceptor submarine project, which was eventually to result in the revolutionary 'Alfa' class, was begun. Some sources state that the decision was made as early as 1956, soon after Krushchev's accession, and that the impetus for the project was provided by the deployment of US submarines armed with the Regulus strategic cruise missile, which had to be launched from the surface.[1] It is conceivable, however, that the project dates from the subsequent major defence policy review of 1959–60, and is more closely related to the anti-carrier programme.

There are similar unresolved disagreements regarding the year in which the first prototype was completed for trials, with some sources stating that it entered service as early as 1967, and others as late as 1972. What is clear, however, is that the development period of the 'Alfa' was protracted, and that the subsequent history of the type was plagued by serious technical problems which have probably been only partially resolved.

A titanium hull

The physical constraints on submarine design are similar in many respects to those which plague the design of surface ships. Higher power generally means greater machinery volume and weight; the direct consequence is a corresponding increase in hull size

Below: *The first production unit of the 'Alfa' class photographed on the surface close to the Kola Peninsula in 1978. Note the 'Albacore' hull-form and low, streamlined conning tower. (MoD)*

and displacement, which in turn brings with it a requirement for higher power. There is therefore a limit to how much additional performance can be squeezed out of a submarine employing conventional construction techniques and a standard machinery installation.

The only way out of this vicious circle is to develop propulsion machinery that has a much higher power output for the same weight and volume, and to employ new lightweight materials for the construction of the hull. For the 'Alfa' the Soviet designers were to use a combination of both. The result was a submarine with performance characteristics that startled both the Western media and Western military Intelligence.

Major economies in hull weight were made by the adoption of titanium in place of steel. Titanium, a relatively scarce and expensive metal, had been used extensively in the aviation industry, but had not previously been employed for the construction of submarines. Its major advantage over the HY80 steel employed for nuclear submarine construction in the West from the early 1960s was that for a given thickness of plate it has 43 per cent greater tensile strength, whereas weight is only 57–58 per cent that of steel. Titanium is also very resistant to corrosion even at high temperatures (hence its use for pipes in many cooling systems) and is virtually non-magnetic.

However, titanium also has a number of properties which make its employment as the skin of a large structure, such as a submarine, problematical. Its modulus of elasticity is only half that of steel; it therefore has a tendency to crack under stress, and has to be alloyed with small quantities of other substances such as aluminium and manganese. It is also difficult to roll and bend, and requires different methods of welding and processing from those employed in conventional steel construction.

Titanium is better when subjected to internal pressures; when exposed to external pressures, as would be the case at deep diving depths, it is subject to compression, bending and shearing. Deformation under stress may be slow, and needs to be carefully monitored. Pressure hull framing systems would have to be specially designed to ensure an even distribution of stress.[2]

The only Western use of titanium in submarine pressure-hull construction has been for small submersibles designed for underwater research or rescue. In 1979 the US Navy's *Alvin* (DSV-2) was fitted with a new titanium pressure hull which increased her maximum diving depth from 2,000 metres to 4,000 metres, and in 1981–2 *Sea Cliff* (DSV-4) was fitted with a titanium operational chamber which increased her diving depth from 2,100 metres to 6,100 metres. However, neither of these craft can stand adequate comparison with the

'ALFA I' CLASS

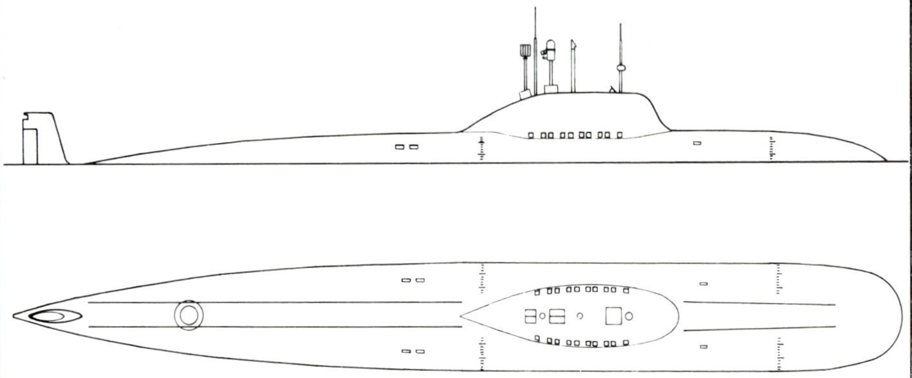

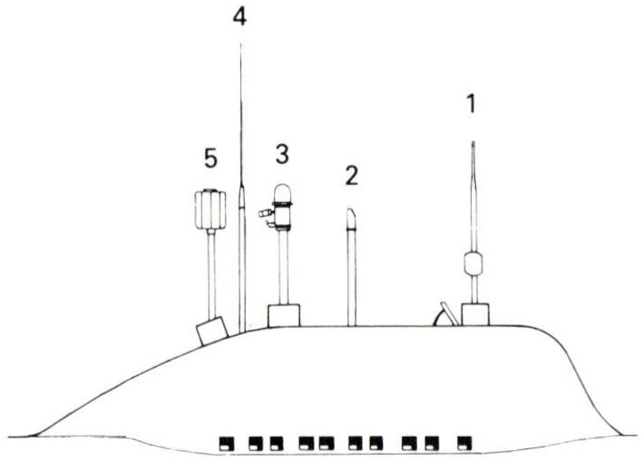

'Alfa'

1. attack periscope
2. search periscope
3. Snoop Head/Bald Head surface surveillance/ESM
4. HF communications mast
5. Park Lamp VLF reception

'ALFA' CLASS

Soviet designation: Podvodnaya Lodka Atomnaya (PLA) (Submarine, Nuclear-Powered)

Construction:
Built at Sudomekh Shipyard, Leningrad; No. 402 Shipyard, Severodvinsk.
Prototype completed Sudomekh 1970; six production units completed 1979–83.

Displacement:
Surfaced: 2,900 tons
Submerged: 3,680 tons

Dimensions:
Length: 81.4m (267ft)
Beam: 9.5m (31ft)
Draught: 7m (23ft)

Propulsion:
Two liquid-metal reactors with turbo-electric drive; one shaft with 5-bladed propeller; 45,000shp for 42 knots.

Complement 45

Weapons:
Six bow tubes of 533mm (21in) diameter for anti-ship/anti-submarine torpedoes and SS-N-15 Starfish ASW missiles? maximum capacity eighteen torpedoes/missiles.

Sensors:
LF active/passive sonar; search and attack periscopes; Snoop Head/Bald Head surface surveillance radar/ESM; Park Lamp DF loop.

Communications:
HF and VHF radio.

Soviet 'Alfa'. In both cases the pressure hull is spherical, not cylindrical, and diameter is a mere 2.4 metres.

The hull-form of the 'Alfa' is very different from that of other contemporary Soviet nuclear boats. Whereas the 'November', the 'Charlie' and the 'Victor' have a 'tear-drop' configuration, with a relatively full bow section, that of the 'Alfa' is more reminiscent of the 'airship' configuration adopted by the US Navy for the experimental *Albacore* and for its mainstream nuclear successors of the *Skipjack* class. It is a hull-form designed to minimize underwater resistance at the cost of bow volume which could be used for more extensive sonar provision and weapons stowage.

The low, streamlined fin has a broadly triangular cross-section, the sides being faired into the outer hull casing in such a way as to eliminate all sharp angles. This serves to improve hydrodynamic performance.

The configuration of the outer hull suggests that the pressure hull is formed, as in the Dutch and US Navy diesel boats which adopted the 'Albacore' hull-form, by a series of cylinders of different diameters, the broadest of the cylinders being located directly beneath the fin, and housing the control centre, the radio offices, and the accommodation spaces.

At the point where the sides of the fin are faired into the hull casing there is a row of square free-flood holes. Large, rectangular apertures are also visible at a lower level forward and aft of the fin.

The outfit of sensor masts is on a par with that of the 'Charlie' and the 'Victor'. There are search and attack periscopes, HF and VHF/UHF radio communications masts, together with a Park Lamp DF loop. However, the functions of surface surveillance and electronic support measures (ESM) are combined in a new mast-

Right: *Overhead of an 'Alfa'-class SSN taken in the Northern Fleet area. The 'Alfa' is significantly smaller than other Soviet nuclear-powered submarines, thanks to her 'dense' propulsion plant. The distinctive mast at the forward end of the conning tower is the attack periscope. (MoD)*

Right: *Close-up of the conning tower. The sensor masts, from left to right are: search periscope, Snoop Head/Bald Head combined surface surveillance radar/ESM, HF communications, and Park Lamp Mod DF loop/VLF reception. (US Navy, 1983)*

mounted antenna designated Snoop Head/Bald Head, comprising a small radome with side plates suspended beneath and a surface surveillance antenna mounted on its face.

The 'Alfa' is credited with six bow torpedo tubes of standard 533mm diameter, and these are disposed in identical fashion with those of the 'Victor', with one pair of tubes superimposed above a bank of four within the upper section of the bow. The active/passive attack sonar is presumably located beneath the tubes. The pressure cylinder forward of the fin houses the torpedo handling room, which because of its restricted size would have a maximum capacity of no more than twelve reloads. This figure would be more than adequate for a submarine designed not for long-distance patrols, but as a high-speed 'interceptor'. The torpedo handling gear is probably highly automated both in order to reduce the workload on the small crew and to ensure rapid reloading.

The after section of the 'Alfa' is occupied almost exclusively by the high-performance propulsion plant, which employs a single shaft to propel the submarine at a top speed estimated by Western Intelligence sources at about 42 knots. A 5-bladed propeller is employed abaft cruciform after control surfaces. In contrast to other Soviet submarines types, the forward hydroplanes are located beneath the surfaced waterline, as in the latest British attack submarines. They are almost certainly retractable, to enable the 'Alfa' to come alongside without damage.

A liquid metal reactor

In order to control the fission process to a level at which usable energy is produced, the fuel rods in a nuclear reactor are surrounded by a 'moderator' which slows the fast-colliding neutrons. Energy (in the form of heat) is then extracted by passing a coolant through the core.

The coolants most commonly used, water and gas, become radioactive as a result of this process. The energy therefore has to be transferred via a heat-exchanger (or steam generator) to a secondary loop which carries steam to and from the turbines (or, if turbo-electric drive is employed, to the turbo-generators). Because the coolant in the primary circuit is radioactive, the heat-exchanger has to be enclosed within the reactor shielding. This makes a substantial contribution to the overall weight of the installation.

Most Soviet, and all current Western nuclear propulsion plants employ water as both moderator and coolant. US and Royal Navy nuclear plants use saturated steam at relatively low temperatures. This enables the heat-exchanger to be reduced in size, and makes

Left: *A striking bow view of an 'Alfa', taken in 1987. Note the broadly triangular cross-section of the conning tower, and the way the plates at its base are faired into the hull structure. The caps of the torpedo tubes, arranged as an upper pair above a bank of four, are just visible above the surfaced waterline. (MoD)*

control more effective over a wide range of operating conditions.[3]

However, low-temperature steam technology effectively increases the size, and therefore the weight, of the turbine machinery itself. This imposes a natural ceiling on the horsepower that a submarine of given dimensions and displacement can produce.

A number of alternatives to pressurized water for naval applications were proposed during the 1950s. The USS *Seawolf* (SSN-575), was laid down in 1953 and completed in March 1957 with a reactor using liquid sodium as a coolant, and graphite as a moderator. The use of liquid metal made possible a significant rise in coolant temperature, which in turn raised the temperature in the steam generator by several hundred degrees. Following a serious leak in the superheater in 1959, however, the plant was condemned as unreliable and was ultimately replaced by a conventional water-cooled reactor similar to the one installed in *Nautilus*.

There were also proposals during the late 1950s for systems employing helium, lead, bismuth and hydrocarbons. According to Norman Friedman[4] a lightweight fast-neutron gas turbine, probably using helium, was under consideration for the US Navy's *Skate* class of 1957–8, and a nuclear propulsion plant using hydrocarbons, intended for use in destroyers, was under development for a time. However, a combination of technical problems and shortage of time for development led the US Navy to abandon both proposals in favour of proven, more conservative designs.

Because development of the Soviet 'Alfa' was embarked upon as a long-term project, lengthy delays in the deployment of the submarine were more acceptable. The desire to achieve very high speeds in a relatively small submarine could only be fulfilled by the adoption of an exceptionally 'dense' propulsion plant, with a high operating temperature and a very high power to weight ratio.

Lead-bismuth was ultimately selected as the coolant for the 'Alfa' reactor. The major advantage of lead-bismuth over the sodium coolant adopted for the *Seawolf* reactor (and subsequently for a number of experimental civil 'fast-breeder' reactors in the West) was that these particular metals did not become radioactive. The steam generators could therefore be located outside the shielding, which was limited to the reactor itself.

As with sodium, a lead-bismuth coolant operates at atmospheric pressure. There is less stress on pipework than in a water-cooled propulsion plant, where the coolant has to be kept at a pressure of up to 150 atmospheres to prevent it from boiling. The reactor structure is therefore significantly lighter than that of a PWR reactor.

However, liquid metal coolants have to be constantly heated in order to prevent them from solidifying, and a lead-bismuth mixture creates considerable demands in this respect. The melting-point for sodium-potassium is about 90 degrees Celsius. Experience with the British Fast Breeder Reactor at Dounreay has established that it can be allowed to solidify when the reactor is shut down, and that a single isolated heat source is sufficient to liquefy the coolant again.

The melting-point of a lead-bismuth mixture is much higher (in excess of 200 degrees), and the coolant must under no circumstances be allowed to solidify. Moreover, a single isolated heat source is insufficient; the coolant has to be heated throughout. It is probable that

Right: *An 'Alfa' running on the surface. The Snoop Head/Bald Head combined surface surveillance/ESM mast is particularly prominent in this view. (MoD, 1987)*

the lead-bismuth coolant in the 'Alfa' propulsion plant has to be constantly heated by steam, even when the submarine is alongside with its reactor closed down.

Unlike other contemporary Soviet submarine types, the 'Alfa' is thought to employ turbo-electric drive, using turbo-generators to supply power to a large electric motor which drives the single 5-bladed propeller.

The propulsion plant is remarkably compact, and it is probable that a high level of automation was adopted for the machinery installation, with the engine room being unmanned under normal conditions. This would further reduce the requirement for reactor shielding.

Technical problems

The 'Alfa' development programme was protracted, and was apparently beset by serious technical difficulties. The prototype, which finally entered service in 1970, was broken up only four years later following extensive trials in the Baltic and the Northern Fleet areas. A serious accident involving the 'freezing' of the liquid-metal coolant seems the most likely possibility, although problems may also have been experienced with the welding of the titanium hull, resulting in leaks and distortion when operating at depth.

Some of these problems appear to have been resolved, because the Sudomekh and Severodvinsk Shipyard subsequently embarked on slow series production of the class, the second unit being completed during the late 1970s. Production was clearly achieved only at high cost, as is indicated by Soviet references to the 'Alfa' as *zolotaya ryba* ('golden fish'), and was finally terminated in the early 1980s after only six units had been completed.

In spite of the impact made by the 'Alfa' in Western military circles, the absence of any lineal successor in the Soviet Navy suggests that the 'Alfa' probably represents something of a 'dead end' in Soviet submarine development. There are indications that the high theoretical performance of the type has not been matched by its reliability in service, and that the 'Alfa' will continue to prove difficult to operate and to maintain.

It is one thing to construct a submarine hull capable of descending to depths estimated to be about 700 metres (i.e., more than twice the maximum theoretical diving depth of a comparable submarine built of high-tensile steel). It is quite another to expect such a hull to withstand the stresses of high-speed operation at such depths over a long period of time, when it is widely accepted that frequent exposure to high under-

water pressures is a key factor in determining submarine service life. The case for titanium as a hull material therefore remains to be proven. Moreover, operating at extreme depths produces additional stresses and demands on fittings such as internal pipework, blowing systems, and torpedo launching systems.

High-technology propulsion plant such as that installed in the 'Alfa' also creates its own demands, both on components and on the crew. Liquid-metal coolants have to be constantly heated in order to avoid freezing, and the high operating temperatures of the propulsion plant will undoubtedly create additional thermal stresses on pipework and other system components. A high level both of training and of vigilance is therefore required of the crew, combined with frequent monitoring of equipment at the shore base.

Some general observations

Technical reservations apart, the factor which perhaps more than any other has been responsible for the termination of the 'Alfa' production line is that the concept of the fast 'interceptor' submarine is no longer as relevant to the maritime situation in the 1980s as it had been when the project was first conceived.

Below: *The high underwater performance of the 'Alfa' is unquestioned, but the 'Alfa' has been a costly experiment, and there are indications that the technological problems experienced with the submarine will outweigh its utility in an operational context. (MoD)*

There is every indication that the 'Alfa' has achieved the high underwater speeds for which it was designed, and that it can operate at greater depths than its Western counterparts. These features have not unnaturally caused considerable concern in the West, as lightweight anti-submarine torpedoes launched from surface escorts or helicopters, such as the US Mk 46, would not only have to descend to great depths in order to home on the submarine, but would then have insufficient speed to catch it. The appearance of the 'Alfa' has therefore been largely responsible for major Western programmes to produce fast, deep-diving anti-submarine torpedoes.

But the relatively small size of the 'Alfa' has effectively precluded arming it with anti-ship stand-off missiles. It would have to fire its torpedoes at relatively close range, much like the obsolescent 'November' and its diesel-powered counterpart, the 'Foxtrot'. This should expose the submarine to a more extensive (and intensive) range of countermeasures than later types such as the 'Charlie' and 'Oscar', which do not need to approach a task force in order to fire their own weapons.

Some sources credit the 'Alfa' with the ability to fire missiles against hostile submarines, but the design is not particularly well suited to anti-submarine warfare. The 'Alfa' is inherently noisy, particularly when operating at the high underwater speeds for which it was designed, and there is almost certainly inadequate bow volume to accommodate a high-performance sonar and six torpedo tubes. It is far more likely that these submarines are fitted with a sonar optimized for the acquisition of surface contacts, such as that fitted in the 'Charlie'.

Finally, the 'Alfa' represents a 'brute force' approach to technology which is indicative of its 1950s origins. It has proved to be a costly experiment which is unlikely to be repeated.

The 'Papa'

The second developmental type to appear at about this time was the 'Papa', a nuclear-powered submarine armed with anti-ship cruise missiles capable of underwater launch. The solitary unit completed was launched at the Krasnaya Sormovo Shipyard, Gorky, in 1968 and entered service two years later.

In concept the 'Papa' was clearly derived from the 'Charlie', which was in series production at the same shipyard, but there were a number of features which marked out the 'Papa' as being of more advanced design.

The 'Papa' is a much larger boat than the 'Charlie'. The hull is some fifteen metres longer, and beam is in

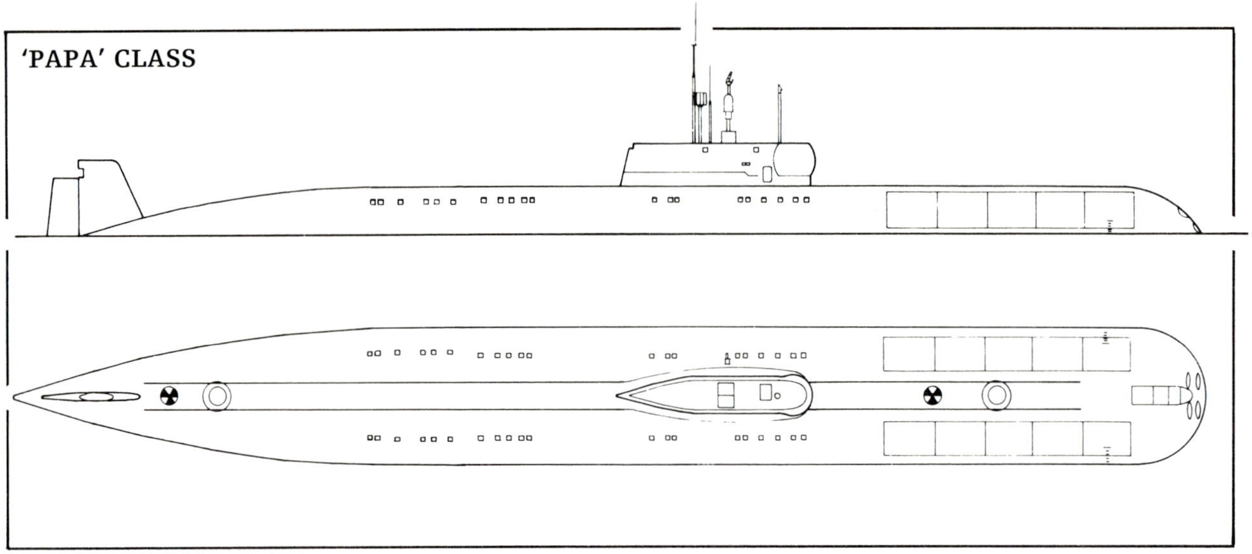

Below: *Unlike its contemporaries of the 'Charlie' class, the 'Papa' has a completely cylindrical hull-form. Close to the bow are the large square hatches covering the launch tubes for the SS-N-9 missiles. Four torpedo tubes are visible above the surfaced waterline, with the torpedo reloading hatch above them. Note the cylindrical bulge at the forward end of the conning tower, which may house an attack sonar. (MoD)*

Right: *Close-up of the conning tower of the 'Papa'. The top of the Brick Pulp ESM antenna is visible between the compass platform and the two communications masts. (US Navy, 1983)*

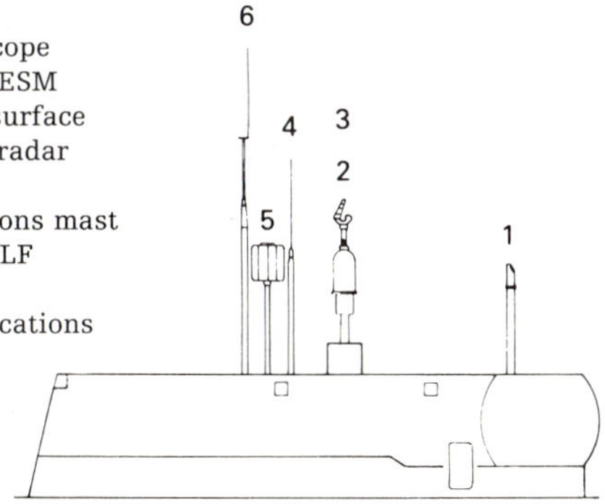

'Papa'
1. search periscope
2. Brick Group ESM
3. Snoop Tray surface surveillance radar
4. VHF/UHF communications mast
5. Park Lamp VLF reception
6. HF communications mast

'PAPA' CLASS

Soviet designation: Podvodnaya Lodka Atomnaya Raketnaya Krylataya (PLARK) (Cruise Missile Submarine, Nuclear-Powered)

Construction:
Built at Krasnaya Sormovo Shipyard, Gorky.
One unit completed 1970.

Displacement:
Surfaced: 6,700 tons
Submerged: 8,000 tons

Dimensions:
Length: 109m (358ft)
Beam: 12m (39ft)
Draught: 9.5m (31ft)

Propulsion:
Two pressurized water-cooled reactors driving geared steam turbines; two shafts with 5-bladed propellers; 60,000shp for 35 knots plus.

Complement 85

Weapons:
Ten SS-N-9 Siren anti-ship missiles in fixed angled launch tubes; six bow tubes of 533mm diameter; anti-ship/anti-submarine torpedoes and SS-N-15 Starfish ASW missiles; maximum capacity fourteen torpedoes/missiles.

Sensors:
LF active/passive sonar; search and attack periscopes; Snoop Tray surface surveillance radar; Brick Pulp/Brick Spit ESM; Park Lamp DF loop.

Communications:
HF and VHF radio.

excess of 12 metres, as compared with only 10 metres for the 'Charlie'. The 'Charlie I' displaces an estimated 4,800 tons submerged; the 'Papa' 8,000 tons.

The hull-form adopted bears only a superficial resemblance to that of the 'Charlie', and is much closer to that of the latest US Navy types. In contrast to the 'tear-drop' configuration of the 'Charlie', the outer casing (and presumably the pressure-hull itself) is cylindrical throughout its length, tapering only at the stern, which is marked by a prominent 'notched' rudder fin. The after control surfaces are cruciform, but the forward hydroplanes are not visible above the surfaced waterline, and are presumably fitted at the mid-point of the hull as in the 'Alfa' and contemporary British SSNs.

The fin is longer and lower than that of the 'Charlie', and has a more sharply angled after edge. The additional length of the fin appears to be related to a prominent bulged fairing approximately 3.5 metres in diameter at its forward end. This presumably houses an active/passive sonar array for target acquisition and fire control. The sensor masts, on the other hand, appear to be standard for the period.

The fully rounded bow section is different from that of the 'Charlie' in that four out of an estimated six 533mm torpedo tubes are visible above the surfaced waterline, and the torpedo reload hatches are located immediately above them. This would preclude the location of a sonar array in the upper section of the bow, but the unusually high position of the tubes suggests that there may be a large low-frequency array in the 'chin' of the submarine.

The missile tubes are beneath large square hatches on each side of the bow section. The hatches are of similar length to those on the 'Charlie', but there are ten of them, as compared with only eight on the earlier boats. The launch tubes are thought to house the SS-N-9 missile, a solid-fuel rocket similar in concept to the SS-N-7. First seen aboard the small rocket ships of the 'Nanuchka' class in 1969, the SS-N-9 (NATO code-name:

Siren) is a larger missile with a length of about 8.85 metres and a range estimated at 60 nautical miles.

In order to take full advantage of this increase in missile range, sonars capable of acquiring the target at the equivalent of two convergence zones would be required. This would account for the enhanced sonar provision evident in the 'Papa' design. The limited precision of such targeting methods, however, would ensure that the 'Papa' remained as dependent on 'off-board' sensors for initial target data input as earlier Soviet submarine types.

Service history

The 'Papa' transferred to the Soviet Northern Fleet shortly after completion, and subsequently embarked on a lengthy series of trials. Since the early 1970s the submarine has undergone two major refits and seen little active service.

The absence of any natural lineal successor has served to confirm the developmental nature of the 'Papa', but the lack of active service, taken together with the amount of time spent in dockyard hands, also indicates that problems may have been experienced with some features of the design.

It has been suggested that the pressure hull, like that of the 'Alfa', is of titanium, not steel. This appears unlikely, given the much greater size of the 'Papa' and the previous association of high-technology submarine developments with the Sudomekh Shipyard.

More likely is that the 'Papa' was designed as a test-bed for a new type of nuclear reactor intended for later classes of missile submarine. The latter would be much larger than the nuclear boats of the 'Yankee'/'Charlie'/'Victor' generation. They would therefore require a reactor of much greater power than the standard 15,000–22,000shp models installed in these types.

The reactor fitted in the 'Papa' is generally credited with a power rating of at least 30,000shp. This is consistent with the theory put forward by some commentators that the reactor was intended for eventual installation in the 'Oscar' class, and possibly also for the 'Typhoon'. There are two reactors in line, and the turbine machinery powers two shafts, giving a theoretical maximum horsepower of about 60,000shp – twice that of the contemporary 'Victor'-class SSN, and three times that of the 'Charlie'. Maximum speed has been estimated at 35 knots, with some sources going even higher.

In spite of its high power rating, the reactor is almost certainly of conventional design, using pressurized water as coolant and moderator. This accounts for the large size of the 'Papa' as compared with other contemporary Soviet attack submarines.

It may be that the problems experienced with the design during trials were related to this new, advanced propulsion plant, which developed horsepower far in excess of any Western submarine of the period. There can be little doubt, however, that the experience acquired by the Soviet Navy in operating the 'Papa' was invaluable in the development of a new generation of nuclear reactors to power a new generation of large submarines.

Below: *The 'Papa' at speed in northern waters. Note the distinctive 'notched' rudder-fin. (MoD)*

6. The Bastion Concept

The year 1972–3 constituted a watershed in Soviet submarine operating concepts. It was the year in which many of the programmes instituted following the political demise of Nikita Krushchev finally bore fruit. Henceforth the Soviet submarine fleet would again assume a primary strategic mission, and all other Soviet naval forces, including major surface units and 'attack' submarines, would be charged primarily with support of that mission.

The reasons for this change of policy have been discussed in an earlier chapter. The value of ballistic missile submarines as a strategic reserve, to be held in readiness against a pre-emptive counterforce strike on Soviet land-based missile silos, was by now fully accepted by the Soviet leadership. The new generation of Soviet SSBNs would not be limited to specifically naval targets, as had the 'Yankees'. On the contrary, they would become a central element in Soviet strategic deterrence, and 'attack against the shore' would supplant 'defence of the homeland' as the primary mission of the Soviet Navy.

The importance of the SSBN force to future Soviet naval policy resulted in a major channelling of resources into new submarine-launched missiles. The primary weakness of earlier Soviet SLBMs had been their limited range, which necessitated lengthy transits by the missile submarines to their designated patrol areas. Of a total of some 65 operational ballistic missile submarines, the Soviet Navy had rarely managed to maintain more than half a dozen on station simultaneously. Moreover, all submarines deployed from the Northern Fleet area to the North Atlantic had to pass through major NATO ASW barriers, while even in the Pacific they were vulnerable to detection by underwater hydrophones which formed part of the increasingly extensive and sophisticated SOSUS ocean surveillance network.

This situation was clearly unacceptable if the SSBN force was to become a key factor in national security. The missiles to be installed in future Soviet submarines would therefore require a range more than three times that of the SS-N-6 which equipped the 'Yankee' class. A range in excess of 4,000 nautical miles would not only obviate the need for Soviet SSBNs to pass through Western anti-submarine barriers, but would enable the submarines to deploy within patrol areas which could be protected by friendly forces. Gorshkov would thereby achieve his goal of supporting his submarine fleet with surface ships and with aircraft, by creating defended 'bastions' in which the Soviet SSBNs could operate freely and without interference.

High-speed operation, which had been considered necessary for rapid transit and for evading enemy forces, would no longer be required. This in turn would reduce the likelihood of detection by hostile submarines and fixed underwater hydrophones belonging to the US SOSUS network. Moreover, the problems experienced in developing accurate inertial navigation systems noted in connection with earlier classes of Soviet ballistic missile submarine would be eased by operating in areas where the mapping of the seabed would be relatively straightforward.

Below: *The map illustrates the fundamental changes in the operational mode of Soviet SSBNs made possible by the SS-N-8 missile. Submarines equipped to fire the latter would no longer need to pass through the NATO anti-submarine barrier in the GIUK Gap in order to reach their patrol areas.*

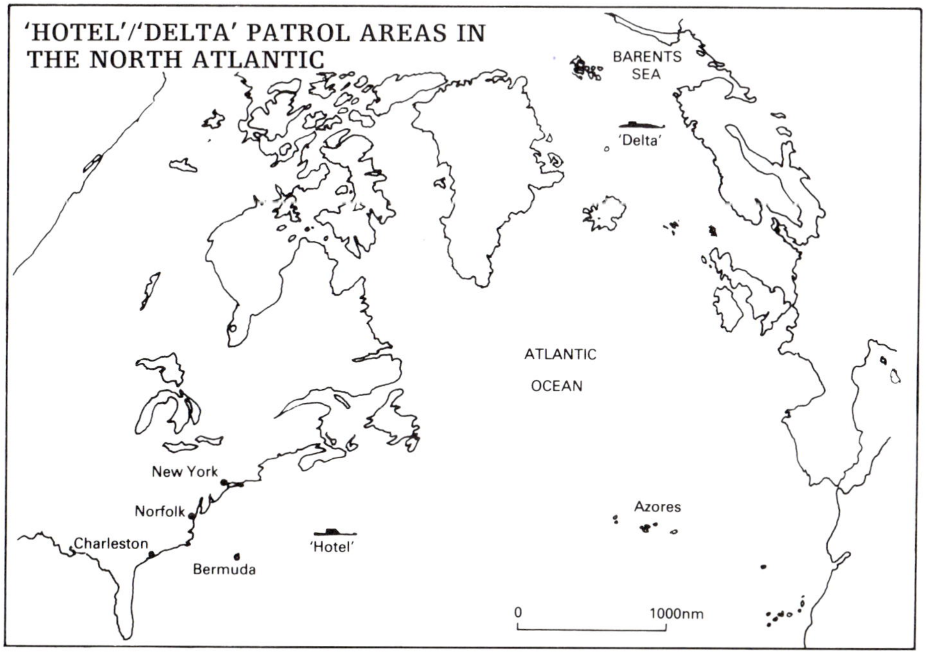

The Bastions

The Barents Sea provided a natural operating area for Soviet SSBNs based with the Northern Fleet. With the

deployment of the 4,200nm SS-N-8 missile (NATO codename: Sawfly), most military and civilian targets on the North American land mass could be covered from these waters, and coverage could be further extended if submarines, supported by Soviet surface forces, were to deploy into the northern reaches of the Norwegian Sea. The line between the Kola Peninsula and Spitzbergen formed the natural perimeter for the northern bastion, and in time the Soviets would create their own network of underwater sensors in this area to support the anti-submarine task groups charged with excluding hostile attack submarines.

The Pacific Ocean itself provided no such natural boundaries. Talinskaya Bay, close to Petropavlovsk on the eastern side of the Kamchatka Peninsula, had for a number of years provided submarines of the Soviet Pacific Fleet with access to open waters. Petropavlovsk was in many ways less than ideal as a military port, being ice-bound for part of the year and isolated from the major Soviet bases on the Sea of Japan by a total absence of road and rail communications; it has to be supplied and supported almost exclusively by air and sea. Nevertheless, it was outside the ring of barriers and 'choke points' which the US Navy, in conjunction with its Far Eastern allies, would attempt to set up

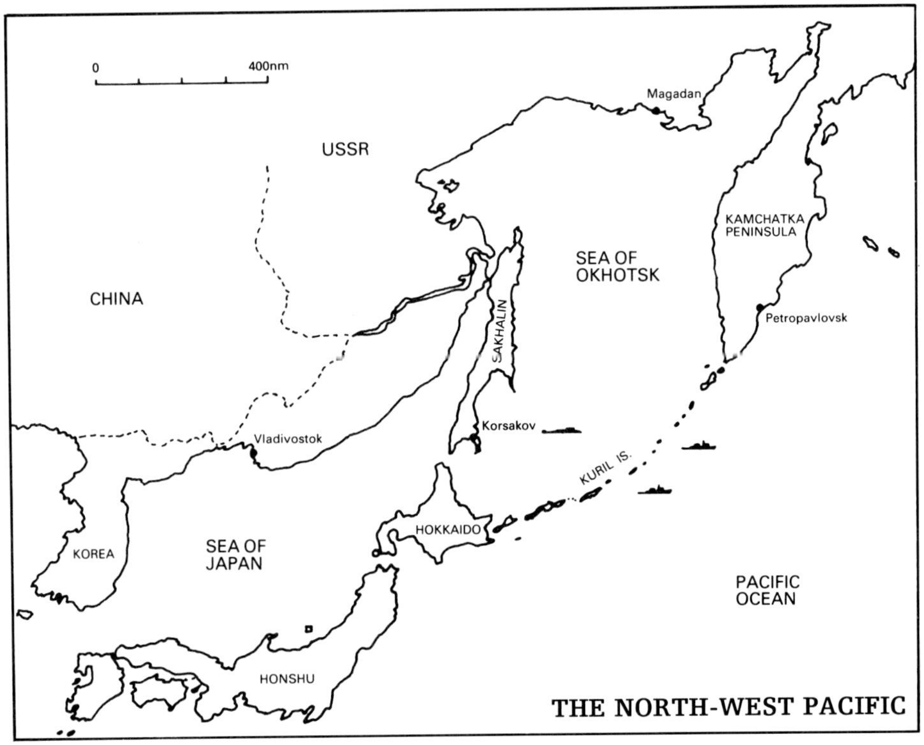

THE NORTH-WEST PACIFIC

Left: *The 'Delta', with its 4,000nm SS-N-8 ballistic missile, made possible a major reorientation of Soviet naval strategy around the 'bastion' concept, whereby Soviet SSBNs were herded together in 'safe' areas and protected by a combination of surface, aerial and submarine forces. (MoD, 1986)*

Left: *The Sea of Okhotsk was to become the Soviet SSBN bastion in the Far East. Submarines entering the Sea either from the Pacific or from the Sea of Japan could conduct their patrols in relative security, while the 'gates' in the Kurile Islands chain would be guarded by anti-submarine vessels and patrol aircraft.*

around the Soviet Fleet based at Vladivostok in the event of hostilities.

The problem with SSBN operations in the open waters of the Pacific was that the United States had taken full advantage of its extensive influence in the area to set up numerous listening-posts, linked to centralized command posts charged with co-ordinating the various data inputs, for a system of underwater hydrophones similar to that which by now ringed the Atlantic. These were located on the seabed off the west coast of the United States, around major Pacific bases such as Hawaii and Guam, and along the chain of islands which borders the northern Pacific, the Aleutians.

The advent of the SS-N-8 missile would make it possible for SSBNs serving with the Soviet Pacific Fleet to operate within patrol areas outside the reach of the US acoustic detection network, and they could be defended with relative ease against incursion by hostile attack submarines.

In accordance with the Yalta Agreement of February 1945, the Soviet Union, in return for a declaration of war against Japan, received the southern half of the island of Sakhalin and the southern islands in the Kurile chain, which stretches from the southern tip of Kamchatka down to the northern tip of Hokkaido. Soviet possession of these islands has remained a source of friction with Japan ever since, but the Soviet Union has shown no inclination to discuss the issue. On the contrary, the importance of these islands to Soviet security in the Far East has increased over the years, and recent Japanese complaints have focused on the 'militarization' of the islands by the Soviets.

It is now generally accepted that the Sea of Okhotsk, which is enclosed by the Kurile chain, is the major operating area for Soviet SSBNs based in the Far East. The islands form a natural outer perimeter, because hostile attack submarines attempting to penetrate the bastion would have to pass through the deep straits which punctuate the chain. These constitute choke points in which Soviet anti-submarine forces, comprising surface ships, aircraft and submarines, could easily be concentrated. Soviet SSBNs can enter the sea either from the Pacific, if based at Petropavlovsk, or via the narrow straits on each side of Sakhalin Island if based at Vladivostok.

The Sea of Okhotsk is a most inhospitable area. The northern half is shallow and icebound, but the southern half has a softer marine climate, and the deep waters (up to 2,500 metres) are particularly well-suited to SSBN patrol operations. Surface navigation is difficult, particularly in the northern waters, where the ice is solid throughout the winter. In the summer there are dense fogs to contend with, caused by warm water from the Pacific meeting the colder waters around the continental land mass. Sonar conditions are poor, because of fluctuations in the thickness of the cold-water layer and because of the continental slopes. These features favour concealment, and make Western anti-submarine counter-operations difficult.

The importance of the Sea of Okhotsk to Soviet strategic operations is evidenced by the efforts made to improve port facilities in the area, notably at Magadan, and by the construction of a second Siberian Railway linking Ustkut and Komsomolsk. The militarization of the Kurile Islands, where a Marine Rifle Division and large numbers of anti-submarine aircraft are now stationed, provides further evidence of the attempt to create a favourable environment for Soviet SSBN operations. The Sea of Okhotsk has also become increasingly important as a testing range for new types of submarine-launched ballistic missile during recent years, with sixteen missiles (nine SS-N-6, four SS-N-8 and three SS-N-18) being launched since 1981.

The 'Delta'

The tendency of Soviet submarine development and weapons development to proceed independently of each other has been noted in an earlier chapter. The revolutionary capabilities of the SS-N-8 missile were not therefore matched by the development of a new type of submarine.

The 'Yankee' construction programme was by now in full swing, and the changeover to a completely new design would have required considerable re-tooling in the shipyards, with adverse consequences for production. Moreover, the Soviet Navy appears to have been generally satisfied with the overall performance of the 'Yankee', which constituted a marked improvement in every respect on previous ballistic missile types.

The problem was that the SS-N-8 missile was so much larger than the SS-N-6 missile around which the 'Yankee' had been designed, that it really required a larger submarine. The increase in missile diameter (0.15m) was not a serious problem, in that it could be compensated by adopting a smaller number of launch tubes, but the increase in length (12.95m for the SS-N-8, as compared with about 11m for the SS-N-6) was more difficult to accommodate, as it is the length of the missile tubes which dictates the dimensions of a ballistic missile submarine's hull.

This problem had long been recognized in the US Navy, where the physical parameters of new missiles have often been determined by the size of the missile launch tubes on existing submarines, and the dimensions of the launch tubes have in turn been determined by those of missiles currently projected. Hence the

'DELTA I' CLASS

Soviet designation: Podvodnaya Lodka Atomnaya Raketnaya Ballisticheskaya (PLARB) (Ballistic Missile Submarine, Nuclear-Powered)

Construction:
Built at Shipyard No. 402, Severodvinsk; Komsomolsk-na-Amur.
Eighteen units completed 1972–7: one in 1972, four in 1973, six in 1974, two in 1975, two in 1976, three in 1977.

Displacement:
Surfaced: 9,000 tons
Submerged: 11,750 tons

Dimensions:
Length: 140m (459ft)
Beam: 12m (39ft)
Draught: 8.7m (29ft)

Propulsion:
Two pressurized water-cooled reactors driving geared steam turbines; two shafts with 5-bladed propellers; 44,000shp for 25 knots max.

Complement 120

Weapons:
Twelve SS-N-8 Sawfly ballistic missiles in vertical launch tubes; six bow tubes of 533mm (21in) diameter for anti-ship/anti-submarine torpedoes; maximum capacity eighteen torpedoes.

Sensors:
LF active/passive sonar; search and attack periscopes; Snoop Tray surface surveillance radar; Brick Pulp/Brick Spit ESM; Park Lamp DF loop; Pert Spring satellite navigation; Cod Eye radio sextant.

Communications:
HF and VHF radio; VLF buoy.

'DELTA I' CLASS

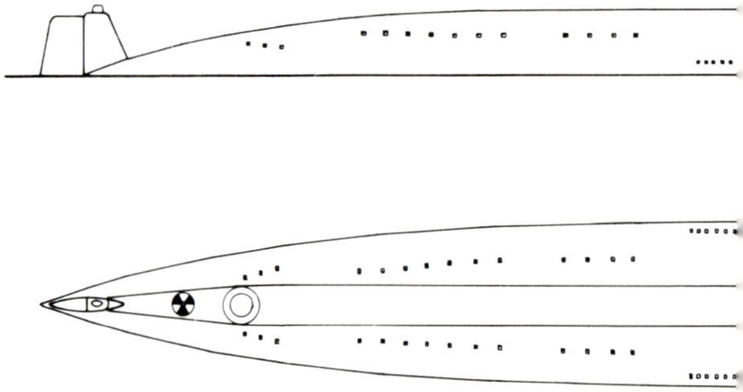

Above: *The 'Delta I' was essentially a modified 'Yankee', with an additional 10m section inserted in the hull, and a higher missile casing. Note the distinctive 'step' between the latter and the after section of the hull. (1976)*

SSBNs of the *George Washington* class went to sea with tubes designed to accommodate not only the initial Polaris A-1 missile, but also its projected successors, the A-2 and A-3. The later *Lafayette* class entered service with the Polaris A-3, but had tubes designed to accommodate the larger Poseidon C-3, which was then under development. The dimensions of the Trident-I (C-4) missile, which was developed for the much larger *Ohio* class, however, were deliberately restricted to enable it to be retro-fitted in the *Lafayette*s. Only when the Trident-II (D-5) missile enters service in the 1990s will the large size of the *Ohio*'s missile tubes be fully utilized.

This close dove-tailing of missile and submarine development is not evident in the Soviet Union, where a different relationship exists between the Soviet Navy and the military-industrial complex. There can be little doubt that the problems experienced by the Navy in acquiring missiles of standard dimensions for its submarines are related to centralized research and development designed to serve all the armed services. Submarine-launched ballistic missiles are not closely

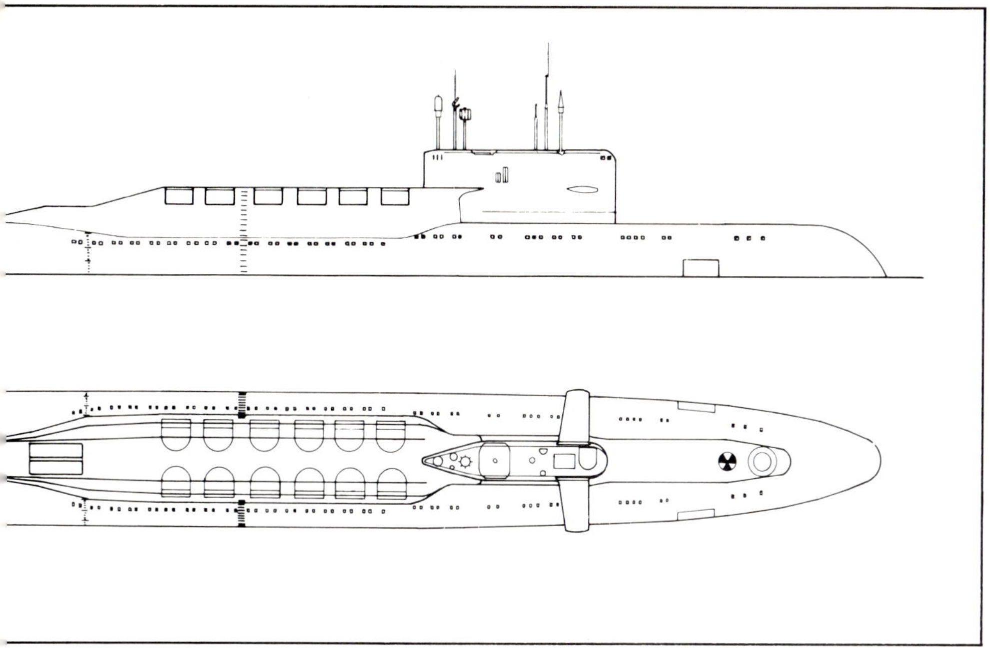

tailored to Soviet Navy requirements, as would be the case in the West, but are a 'spin-off' from land-based missiles, which are not constrained by restrictions on dimensions or by the same safety considerations (hence the Soviet Navy's willingness to accept liquid-fuel SLBMs). In accordance with general Soviet design practice, a 'base-line' model is designed for maximum performance, and is then subjected to successive incremental improvements. Some of these incremental improvements (increased range, for example, or multiple re-entry vehicles) may result in increases in the dimensions of the missile itself, as in the case of the SS-N-18 which was to succeed the SS-N-8.

Since it was considered undesirable to disturb shipyard production, the new 'Delta'-class SSBN would be a modified 'Yankee', with the number of missile tubes reduced from sixteen to twelve, and the height of the deck casing in the vicinity of the missile compartment raised by approximately 1.75 metres.

Even with twelve tubes, the missile compartment of the 'Delta' is some two metres longer than that of the 'Yankee'. One feature that is responsible for this is that

Above: *The twin doors set into the after section of the missile casing are for a VLF communications buoy. The only modification to the 'Yankee' sensor outfit is the addition of a Pert Spring satellite navigation mast.*

Right: *Stern quarter view of a 'Delta I'. The Pert Spring satellite navigation antenna is in the raised position. The mast at the after end of the conning tower is for VHF/UHF communications. (MoD)*

the gap between the fourth and fifth pairs of missile tubes is larger than the gap between the other pairs. The reason for this is not clear, although it may be related to the layout of the missile control panels between decks, as this feature has been retained in all the later 'Delta' variants.

The after end of the missile casing descends in graduated steps to join the after part of the hull casing. An additional 8m hull section was inserted into the submarines at this point. Double hinged doors, some 4.75m long and 2.5m wide, are set into the upper part of the first 'step'. These are thought to conceal a towed buoy for Very Low Frequency (VLF) communications. This enables the submarine to receive radio messages while operating at depth (typically 40 metres), without needing to put a mast-mounted aerial above the surface. The 'Delta' appears to have been the first Soviet ballistic missile submarine to be fitted for VLF communications from the outset, although similar facilities were soon to be extended to earlier classes, including the diesel-powered 'Golfs'.

The after (55m) section of the 'Delta' appears to be identical with that of the 'Yankee', suggesting that the machinery installation was largely unchanged. Given that the 'Delta I' is 10 metres longer than the 'Yankee', and has a submerged displacement estimated at 11,750 tons – more than 2,000 tons greater than that of the 'Yankee' – there must be some reduction in maximum speed. This was presumably acceptable in view of the very different operational mode made possible by the long-range SS-N-8 missile.

Few further modifications were made to the basic 'Yankee' design. The long, slotted, free-flood apertures which lined the hull of the latter were replaced by a larger number of small square apertures, arranged in pairs abeam the missile compartment and the fin. The sonar 'window' was also less prominent, because these submarines were coated with a sonar-absorbent rubber sheathing from the outset.

The only visible modification to the sensor outfit was the addition of a cone-shaped satellite communications aerial, designated Pert Spring, located atop a retractable mast towards the forward end of the fin, on the starboard side.

'DELTA II' CLASS

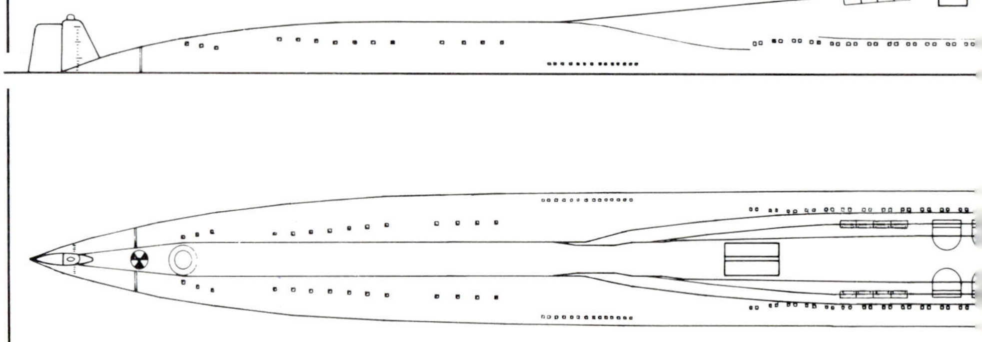

Left: *The 'Delta II' differs from the 'Delta I' in having two additional pairs of missile tubes, for a total of sixteen SS-N-8 missiles. Note the way the missile 'hump' is angled down on to the after casing without a break. The VLF doors are set into the angled section. (MoD)*

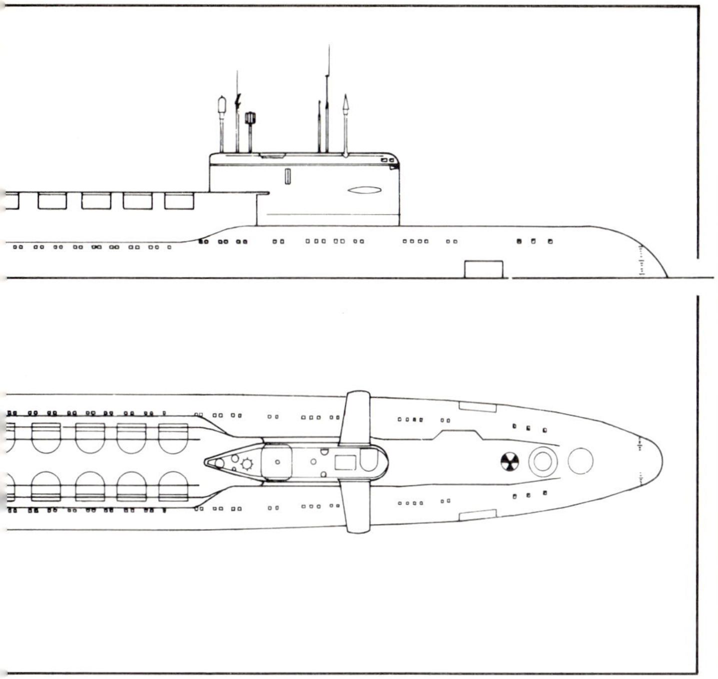

'DELTA II' CLASS

Soviet designation: Podvodnaya Lodka Atomnaya Raketnaya Ballisticheskaya (PLARB) (Ballistic Missile Submarine, Nuclear-Powered)

Construction:
Built at Shipyard No. 402, Severodvinsk. Four units completed 1974–5: one in 1974, three in 1975.

Displacement:
Surfaced: 10,000 tons
Submerged: 12,750 tons

Dimensions:
Length: 155m (508ft)
Beam: 12m (39ft)
Draught: 8.7m (29ft)

Propulsion:
Two pressurized water-cooled reactors driving geared steam turbines; two shafts with 5-bladed propellers; 44,000shp for 24 knots max.

Complement 120

Weapons:
Sixteen SS-N-8 Sawfly ballistic missiles in vertical launch tubes; six bow tubes of 533mm (21in) diameter for anti-ship/anti-submarine torpedoes; maximum capacity eighteen torpedoes.

Sensors:
LF active/passive sonar; search and attack periscopes; Snoop Tray surface surveillance radar; Brick Pulp/Brick Spit ESM; Park Lamp DF loop; Pert Spring satellite navigation; Cod Eye radio sextant.

Communications:
HF and VHF radio; VLF buoy.

'Delta II and III'

In 1974, only two years after the completion of the first 'Delta', a new variant appeared. The 'Delta II' featured an additional 15m section amidships which enabled the submarine to carry sixteen SS-N-8 missiles instead of twelve. Only four submarines of this type were completed, and they were built exclusively at the Severodvinsk Shipyard. Komsomolsk, the second of the SSBN building yards, had by now completed its 'Yankee' programme, and had embarked on construction of the 'Delta I'.

In the 'Delta II' the prominent casing abaft the missile tubes was angled down on to the after hull casing without a break. In addition to the pair of VLF doors, there were now four pairs of smaller hatches towards the top of the angled part of the casing, presumably for access to equipment housed within the 'hump'. Otherwise there were no significant changes either in the submarine's configuration or in equipment.

More significant was the appearance of the first 'Delta III', also built at Severodvinsk, in 1975. The 'Delta III' was intended as a launch vehicle for the SS-N-18 missile, a development of the SS-N-8 which equipped the 'Delta I' and 'Delta II'.

The SS-N-18 was the first operational Soviet SLBM to be designed from the outset to carry multiple, independently targeted re-entry vehicles (MIRV). The Mod 1 variant of the missile carried three MIRVs, each with a 200KT nuclear warhead. Range was correspondingly reduced to about 3,500 nautical miles (i.e., some 700nm less than the original variant of the SS-N-8). Development of the SS-N-18 missile would continue, however, and newer, more powerful variants would emerge. The single-warhead Mod 2, introduced shortly afterwards, had a maximum range estimated at 4,350 nautical miles, while the Mod 3, with a range similar to that of the Mod 1, would carry no fewer than seven independently targeted warheads.

The diameter of the SS-N-18 missile was identical with that of the SS-N-8, so no increase in the length of the submarine was necessary over the 16-missile 'Delta II'; but the SS-N-18 was significantly longer than its predecessor, and this was again to have serious implications for the height of the missile casing, which had to be raised a further 1.25 metres in the 'Delta III'. The missile casing was by now fully 9m above the surfaced waterline, as compared with 7.75m for the 'Delta I' and only 6m for the 'Yankee', bringing it almost to the top edge of the fin.

This modification appears to have been possible only at the expense of the handling qualities of the submarine. During surface operations the captain has his view of the after control surfaces and access hatches obstructed. This must create problems when the submarine is coming alongside, particularly as the high length-to-beam ratio of these boats makes them potentially difficult to manoeuvre. Increasing the height of the missile casing must also have had an adverse effect on stability, both in the surfaced and in the submerged

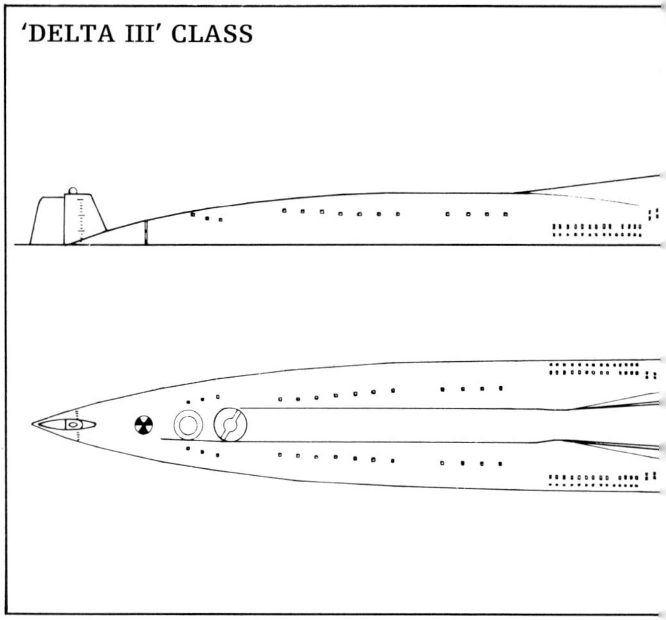

condition. Modifications aimed at correcting these defects were to be incorporated in a later 'Delta IV' variant.

The additional volume created between the outer hull casing and the inner pressure hull by increasing the height of the missile compartment necessitated a major revision of the number and layout of free-flood holes. Three rows of square apertures run the entire length of the missile casing.

Other modifications include the provision of two sets of doors for VLF buoys in the angled part of the missile casing, and the provision of a mating hatch for a deep submergence rescue vehicle (DSRV) on the after part of the hull casing. Earlier types of Soviet submarine had only conventional escape towers beneath watertight hatches at the forward and after ends of the boat.

Construction and deployment

Ten units of the 'Delta III' class were completed at Severodvinsk from 1975 to 1978. The production rate subsequently decreased dramatically, with only one boat per year being completed from 1979 to 1982.

The fall-off in production indicates that the emergency SSBN construction programme initiated during the late 1960s was now drawing to a close. The eleven-year period from 1967 to 1978 had witnessed the completion of no fewer than 34 SSBNs of the 'Yankee' class and 30 SSBNs of the 'Delta' class. This compares with a US Navy figure of 41 SSBNs operational throughout the period. The Strategic Arms Limitation Treaties, SALT I and SALT II, would make further expansion of

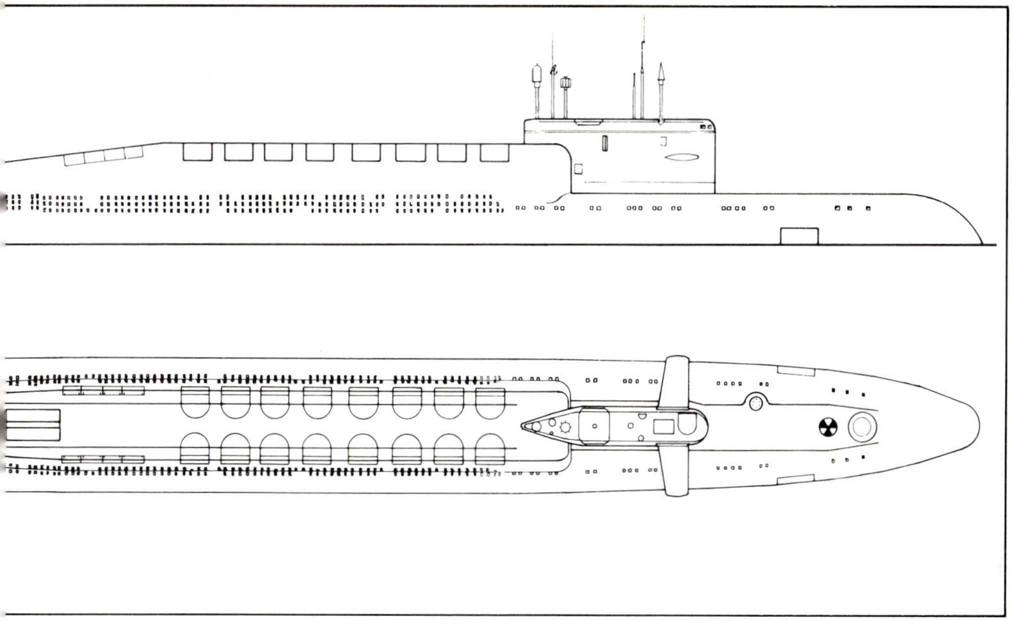

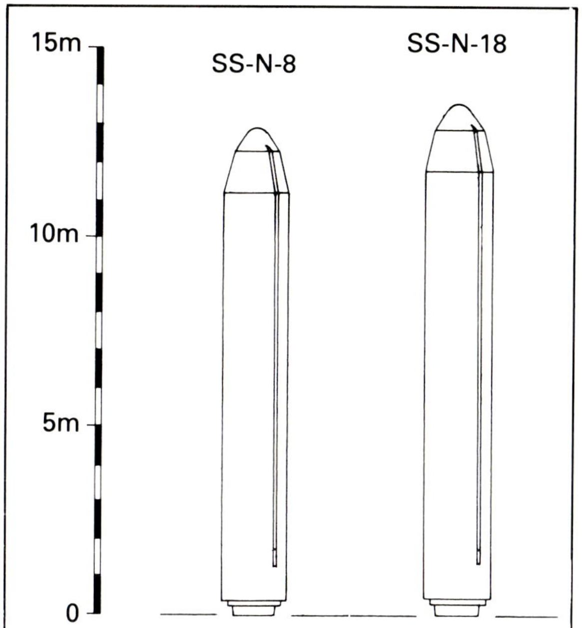

SS-N-8 SAWFLY

Length:	13m
Diameter:	1.8m
Propulsion:	2-stage liquid-fuel rocket
Guidance:	inertial
Range:	Mod. 1: 4,200nm (7,800km)
	Mod. 2: 4,900nm (9,100km)
Warhead:	Mod. 1: 1 RV × 1.2MT nuclear
	Mod. 2: 2 MRV × 800KT nuclear
In Service:	Mod. 1: 1973
	Mod. 2: 1977

SS-N-18 STINGRAY

Length:	13.6m
Diameter:	1.8m
Propulsion:	2-stage liquid-fuel rocket
Guidance:	inertial
Range:	Mod. 1/3: 3,500nm (6,500km)
	Mod. 2: 4,350nm (8,000km)
Warhead:	Mod. 1: 3 MIRV × 200KT nuclear
	Mod. 2: 1 RV × 450KT nuclear
	Mod. 3: 7 MIRV × 100KT nuclear
In Service:	Mod. 1: 1978

Opposite page: *Two views of a submarine of the 'Delta II' class taken in July 1986 in northern waters. Sensor masts, from forward to aft, are: Pert Spring satellite navigation, HF radio communications, Shot Gun VHF/UHF radio communications (starboard and Snoop Tray surface surveillance radar (port). (MoD)*

Left: *The height of the missile casing is particularly evident in this view of a 'Delta III'. Note the rows of hull apertures necessary for rapid flooding of this massive structure, to enable the submarine to dive quickly if caught on the surface by a patrol aircraft. (MoD, 1986)*

the Soviet SSBN force unnecessary. Indeed, the upper limits on sea-launched ballistic missiles established by the latter treaty would result in the decommissioning of first the elderly 'Hotel' class, and subsequently a number of the earlier 'Yankees', to allow the completion of new SSBNs armed with longer-range missiles.

Both the 'Delta II' and the 'Delta III' were built exclusively at Severodvinsk, while the Komsomolsk Shipyard in the Far East continued to build 'Delta I's until construction ceased in 1977. In all, eighteen 'Delta I's were built of which half were completed at Komsomolsk and entered service with the Pacific Fleet. The four 'Delta II's remained in the Northern Fleet area, but seven of the fourteen 'Delta III's were transferred from the Arctic to the Pacific from 1978 onwards.

Of the Pacific 'Deltas', 12–13 boats are currently based at Petropavlovsk, on the Kamchatka Peninsula, and conduct regular patrols in the Sea of Okhotsk. The remaining 3–4 boats of the class are based at Vladivostok, from where they sustain a continuous single-submarine patrol in the Sea of Japan. The squadron at Petropavlovsk also includes seven 'Yankees', which generally conduct their patrols in the north-eastern

'DELTA III' CLASS

Soviet designation: Podvodnaya Lodka Atomnaya Raketnaya Ballisticheskaya (PLARB) (Ballistic Missile Submarine, Nuclear-Powered)

Construction:
Built at Shipyard No. 402, Severodvinsk.
Fourteen units completed 1975–82: two in 1975, four in 1976, two in 1977, two in 1978, one in 1979, one in 1980, one in 1981, one in 1982.

Displacement:
Surfaced: 10,500 tons
Submerged: 13,250 tons

Dimensions:
Length: 155m (508ft)
Beam: 12m (39ft)
Draught: 8.7m (29ft)

Propulsion:
Two pressurized water-cooled reactors driving geared steam turbines; two shafts with 5-bladed propellers; 44,000shp for 24 knots max.

Complement 120

Weapons:
Sixteen SS-N-18 Stingray ballistic missiles in vertical launch tubes; six bow tubes of 533mm (21in) diameter for anti-ship/anti-submarine torpedoes; maximum capacity eighteen torpedoes.

Sensors:
LF active/passive sonar; search and attack periscopes; Snoop Tray surface surveillance radar; Brick Pulp/Brick Spit ESM; Park Lamp DF loop; Pert Spring satellite navigation; Cod Eye radio sextant.

Communications:
HF and VHF radio; 2 VLF buoys.

SOVIET ANTI-SUBMARINE OPERATIONS

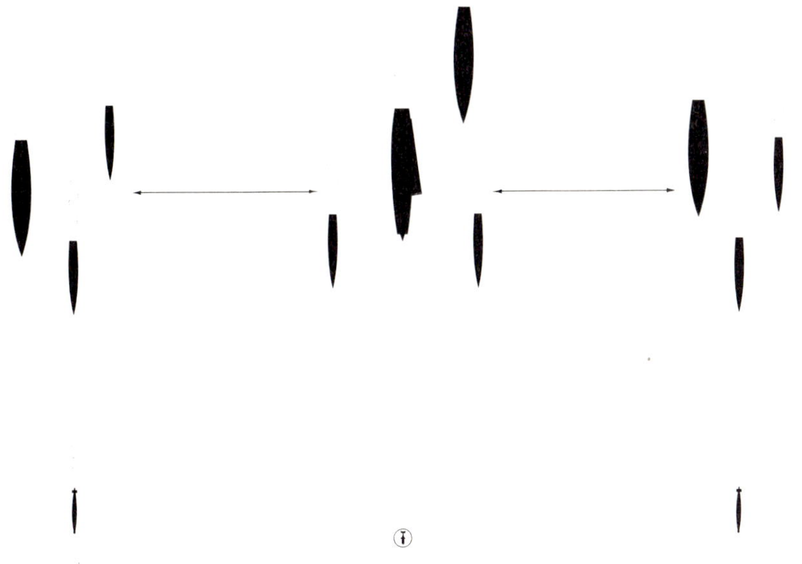

quadrant of the Pacific. Two further 'Yankees' are based at Vladivostok; together with the six remaining SSBs of the 'Golf II' class, they constitute an important element in the Soviet theatre nuclear forces in that region.

'Victor II'

The bastion concept added a new dimension to Soviet anti-submarine warfare. Previously, Soviet ASW had been essentially 'offensive' in nature. The primary mission of major Soviet anti-submarine units was to hunt down, either singly or in groups, Western ballistic missile submarines patrolling in open waters. Now these same anti-submarine units would be tasked with defending the perimeters of the Soviets' own SSBN bastions in the Northern and Pacific Fleet areas.

The defensive mission required different qualities. High speed, which was invaluable in open-ocean operations involving lengthy transits, and which was useful both for closing a potential target and for escaping the consequences of an attack, declined in importance. Quietness, on the other hand, would be essential if

Left: *A 'Delta III' photographed in northern waters in 1986. This variant is identical in length with the 'Delta II', but the height of the missile casing has been raised to accommodate the SS-N-18 missile. Note the twin VLF doors set into the angled section, and the missing hull tiles abeam the conning tower.*

Above: *Soviet anti-submarine operations: Soviet operations in support of the bastions would probably involve small 'task groups' of anti-submarine surface units deployed in a line across the projected path of hostile submarines, and supported by submarines and maritime, patrol aircraft. Overall control of these operations would be exercised from a command post based on land, but tactical control would be delegated to an anti-submarine cruiser of the* Kiev *class.*

Right: *The 'Victor II' has a longer bow section than its predecessor, and was designed from the outset to fire the SS-N-15. (MoD)*

Western attack submarines attempting to enter the bastion were to be intercepted and 'killed'. The task of silencing Soviet SSNs to a point at which the noise generated by their propulsion machinery approached the low levels achieved by contemporary Western attack boats would be a long-drawn-out process. However, Soviet submarines operating in support of the bastions would at least be able to patrol at low speed as they lay in wait for the Western SSNs. Turbine machinery noise and flow noise would therefore be much reduced, although noise from auxiliary machinery such as coolant pumps would remain a problem. Moreover, the Soviet boats would no longer be required to pass through Western ASW barriers in order to perform their essentially defensive mission.

The 'Victor', which had been designed for the anti-SSBN mission, was less than ideally suited to defensive, 'sit-and-wait' tactics. Its two-reactor propulsion plant was designed for high performance, and its sonar, although very capable by Soviet standards, was almost certainly not in the same class as the BQQ-2 system of its US contemporaries. Production rate was also a problem, the Admiralty Shipyard being capable of completing a maximum of two boats per year.

A more capable variant, the 'Victor II', entered service in 1972, but the available evidence suggests that this type was also intended for 'offensive' ASW, and that its primary targets were Western SSBNs. The 'Victor II' was some 7 metres longer than the 'Victor I', the additional length being apparently associated with additional fire control electronics and weapons capacity to enable the new submarine to handle the SS-N-15 missile.

The SS-N-15 (NATO code-name: Starfish) was the Soviet equivalent of the US SUBROC missile, which first entered service in the early 1960s. SUBROC was a solid-fuel rocket which carried a nuclear depth bomb out to a distance of some 30 nautical miles and was launched from a standard 533mm torpedo tube. The Soviet missile is thought to be similar both in overall configuration and in performance, although some degree of backwardness in solid-fuel propulsion technology is evident in the reduced range of the missile, estimated at 20nm. Even this figure would have been more than adequate to match Soviet sonar performance of the day, and the submarine would presumably require target inputs from other supporting ASW units if the range of the missile were to be utilized to the full.

The nuclear payload of the SS-N-15 is probably indicative of an anti-SSBN mission. Even by the early 1970s it was becoming apparent to Soviet military strategists that sea-based nuclear weapons would probably be usable only in the context of all-out nuclear war on land. The FRAS-1 nuclear ASW missile

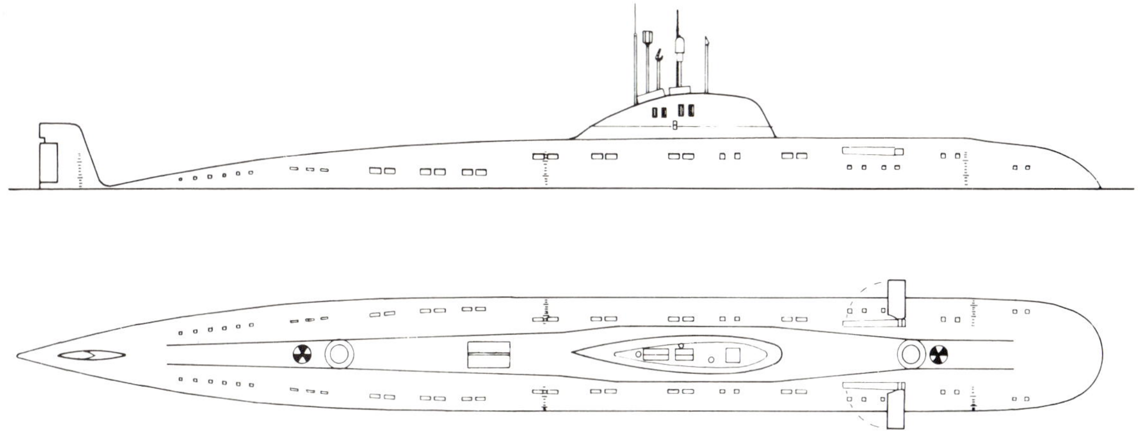

'VICTOR II' CLASS

Left: *Bow view of a 'Victor II'. Note the continuous band of rubberized sheeting around the forward edge of the fin. In the absence of the bridge windows standard on Soviet submarines of the 'HEN' generation, it seems logical to conclude that a sonar array may be located within the forward section of the conning tower. Sensor masts in the raised position are, from left to right: search periscope, HF radio communications, and Shot Gun VHF/UHF communications. (US Navy, 1986)*

Right: *Close-up of the conning tower of a 'Victor II'. Sensor masts, from left to right, are: search periscope, Shot Gun VHF/UHF communications, Park Lamp VLF reception, and HF communications. (US Navy, 1986)*

'VICTOR II' CLASS

Soviet designation: Podvodnaya Lodka Atomnaya (PLA) (Submarine, Nuclear-Powered)

Construction:
Built at Admiralty Shipyard, Leningrad. Seven units completed 1972–8.

Displacement:
Surfaced: 4,500 tons
Submerged: 5,700 tons

Dimensions:
Length: 102m (334ft)
Beam: 10m (33ft)
Draught: 7m (23ft)

Propulsion:
Two pressurized water-cooled reactors driving geared steam turbines; one shaft with 5-bladed propeller; two small 2-bladed propellers on stern planes for slow-speed operations; 30,000shp for 30 knots max.

Complement 80

Weapons:
Six bow tubes of 533mm (21in) diameter for anti-ship/anti-submarine torpedoes and SS-N-15 Starfish ASW missiles; maximum capacity twenty-four torpedoes/missiles.

Sensors:
LF active/passive sonar; search and attack periscopes; Snoop Tray surface surveillance radar; Brick Pulp/Brick Spit ESM; Park Lamp DF loop.

Communications:
HF and VHF radio; VLF buoy.

had been introduced on Soviet surface ships as early as 1967, and was now being complemented (and generally superseded) by the SS-N-14 missile, which carried a homing torpedo with a conventional HE warhead. The development of a similar submarine-launched weapon, the SS-N-16, appears to have been initiated only in the early 1970s, probably in answer to the different tactical requirements demanded by the bastion concept.

Apart from its longer bow section, there were few external differences between the 'Victor II' and its immediate predecessor. As with the 'Delta', there were now twin doors in the after deck casing concealing a VLF communications buoy. Later units of the class also featured new HF and VHF/UHF radio masts (NATO designation: Shot Gun) on which the aerial was encased in a protective sheath.

Seven units of the 'Victor II' class were completed at the Admiralty Shipyard from 1972 to 1978. Somewhat unusually, the early units appear to have been built in parallel with the 'Victor I', which continued in production at the yard until 1975.

Right: *The raised bow casing is particularly evident in this starboard quarter view. Note the positioning of the forward hydroplanes close to the bow. (US Navy, 1985)*

Left: *The 'Tango' is exceptionally large for a diesel-powered submarine. Unlike its immediate predecessors of the 'Foxtrot' class, the 'Tango' has a broad, cylindrical hull-form to enable it to accommodate similar weapons systems to contemporary Soviet nuclear types. Note the slotted free-flood apertures in the hull casing. (MoD, 1987)*

'Tango'

The submarine type which in many respects constituted the first real Soviet response to the demands of the bastion concept was the 'Tango', a diesel-electric boat which first appeared in 1972.

Of similar size to the 'Foxtrot', which had ceased production for the Soviet Navy in the late 1960s, the 'Tango' was at first touted as being its lineal successor. Despite a number of technical similarities, however, the two types were apparently designed with very different missions in view.

The bottle-neck in SSN production created by the scale of the ballistic missile submarine programme meant that the number of 'Victors' that could be completed in the foreseeable future would be inadequate to perform effectively both the 'offensive' anti-SSBN mission and the defensive mission in support of the bastions.

On the other hand, the comparative proximity of the bastion perimeters to Soviet naval bases meant that the long endurance conferred on a submarine by nuclear propulsion was no longer essential. Moreover, diesel-powered submarines were particularly well-suited to low-speed patrols in support of the bastions, and would be capable of near-silent operation when using their electric motors. Diesel-powered submarines utilized existing technology and shipyard tooling, and could therefore be built without prejudice to the nuclear submarine construction programme. If the submarine were sufficiently large, there was no reason why it should not be able to accommodate the latest weapons and sensors designed for the nuclear boats.

The 'Tango' met all of these requirements. Overall length was virtually identical with that of the 'Foxtrot', but beam was increased from 7 metres to 9 metres. More significantly, the traditional 'cigar'-shaped hull-form of early Soviet post-war submarines was abandoned in favour of a more modern cylindrical configuration. This had the effect of increasing internal voume, particularly in the area of the bow, which was fully rounded as in the latest nuclear types. It must also have simplified internal arrangements, as the pressure hull appears to be of constant diameter throughout the length of the submarine.

Not only was the bow fully rounded, but it featured a raised deck casing which serves to distinguish the 'Tango' from earlier types when seen in profile. Early speculation favoured the view that this bow 'hump' was weapons-related and, more specifically, that it was intended to provide additional stowage capacity to enable the submarine to carry the SS-N-15 nuclear anti-submarine missile. This argument is difficult to sustain, however, because it implies that the diameter of the pressure hull was increased at this point, which is clearly not the case.

The rationale for increasing the height of the bow casing appears to be sensor-, not weapons-related. The fully rounded bow of the 'Tango' is broad enough to accommodate the 'crown' hydrophone array of the 'Victor's low-frequency sonar; but the reduced beam of the 'Tango' makes it unlikely that the six bow tubes are disposed in similar fashion to the 'Victor', with a bank

'TANGO' CLASS

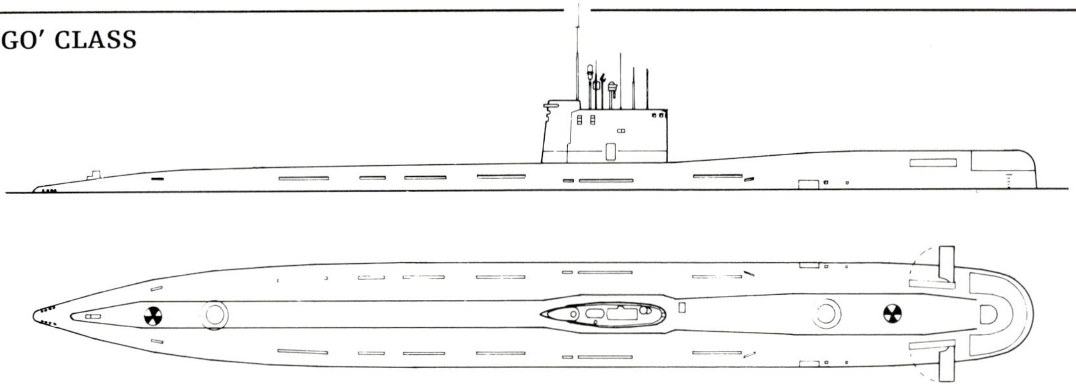

Right: *Close-up of the conning tower of a submarine of the 'Tango' class. Like its nuclear-powered counterparts of the 'Victor' class, the 'Tango' appears to have a rubberized hull coating of anechoic tiles, three of which are clearly missing in this photograph. Masts in the raised position are, from left to right: snorkel, Quad Loop DF antenna, VHF/UHF communications, Squid Head ESM, and HF communications. (1984)*

'TANGO' CLASS

Soviet designation: Podvodnaya Lodka (PL) (Submarine)

Construction:
Built at Krasnaya Sormovo Shipyard, Gorky.
Eighteen units completed 1972–82: two per year.

Displacement:
Surfaced: 3,100 tons
Submerged: 3,900 tons

Dimensions:
Length: 92m (302ft)
Beam: 9m (30ft)
Draught: 7m (23ft)

Propulsion:
Diesel-electric propulsion on two shafts; three diesels, 6,000bhp for 18 knots surfaced; two electric motors, 6,000hp for 16 knots submerged.

Complement 72

Weapons:
Six bow tubes of 533mm (21in) diameter for anti-ship/anti-submarine torpedoes and SS-N-15 Starfish ASW missiles; maximum capacity eighteen torpedoes/missiles.

Sensors:
LF active/passive sonar; search and attack periscopes; Snoop Tray surface surveillance radar; Stop Light or Squid Head ESM; Quad Loop DF.

Communications:
HF and VHF radio.

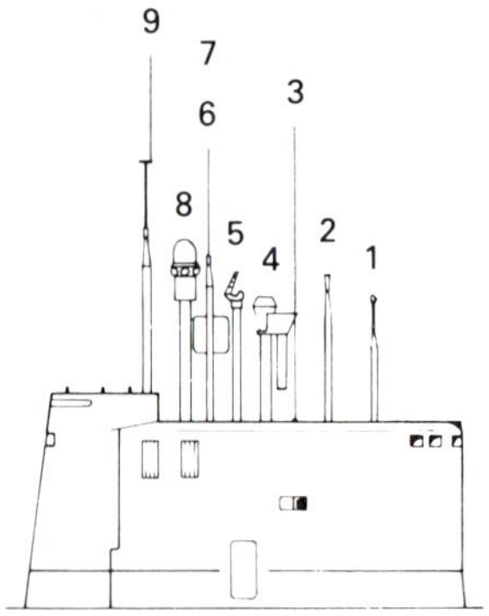

'Tango'
1. attack periscope
2. search periscope
3. HF communications whip
4. snorkel
5. Snoop Tray surface surveillance radar
6. Quad Loop DF
7. VHF/UHF communications mast
8. Squid Head ESM*
9. HF communications mast

*Stop Light ESM in early units

Left: *Although clearly derived from that of the 'Foxtrot', the conning tower of the 'Tango' has an altogether squarer configuration, which has been repeated in the later 'Kilo'. (US Navy, 1976)*

of four tubs immediately beneath the upper pair. The more conventional arrangement of three pairs of tubes disposed one above the other, as in earlier Soviet diesel boats, appears more plausible. The heightened bow casing is therefore necessary to enable the 'Tango' to incorporate the large bow sonar array above the estimated six torpedo tubes. It is also significant that the bow hydroplanes have been placed exceptionally close to the bow, so that they retract into the 'hump' directly abaft the sonar 'window'. This ensures that the pressure hull cylinder can be carried all the way forward at its full diameter. It is the adoption of a cylindrical pressure hull, not the bow 'hump' itself, that is responsible for any increase in weapon stowage capacity.

The combination of increased beam and a modern, cylindrical hull-form, makes the 'Tango' a much larger boat than the 'Foxtrot'. Despite the similarity in length, the 'Tango' displaces fully 3,900 tons submerged, as compared with only 2,400 tons for its predecessor. The largest Western diesel-powered submarines are generally less than 70 metres long, and have a submerged displacement similar to that of the 'Foxtrot'. This provides an indication of Soviet willingness to approach the size of a nuclear-powered submarine in order to incorporate the necessary military qualities into the 'Tango' design.

In spite of this increase in size, however, the diesel propulsion installation has apparently remained unchanged. The three diesel engines are each rated at 2,000bhp according to Western reference sources. The surface speed of 18 knots with which the 'Tango' has been credited therefore seems unlikely, given that the same sources credit the 'Foxtrot', a smaller boat with a hull-form much better suited to surface running, with a maximum speed of only 16 knots.

Standard reference sources also state that there are three shafts. This assessment appears to be based on two assumptions: first, that the propulsion system of the 'Tango' is derived from that of the 'Foxtrot'; and secondly, that the Soviets were still firmly wedded to direct drive, and that three diesels could therefore be equated with three shafts.

The case for both these assumptions remains to be proven. Whereas the 'Foxtrot' was simply a reworking of the basic 'Zulu' design, which is known to have employed direct drive on three shafts, the 'Tango' owes far less to the 'Foxtrot' than is generally believed. Indeed, the hull-form is in many respects closer to that of the 'Juliett', which has only two shafts and was built at the same Gorky Shipyard. Moreover, Soviet persistence with direct drive seems odd, given that diesel-electric drive had been a feature of all Western diesel-powered submarines since the early 1950s. Diesel-electric drive has two major advantages over direct drive: it makes possible a major increase in battery capacity, and therefore in underwater speed and endurance; and because the diesel units drive alternators, not propeller shafts, shafting arrangements are more flexible.

Reference sources agree that the 'Tango' has far greater battery capacity than earlier types. From that point it is but one small step to suggest that the three diesels power two shafts via two – not three – electric motors. The most recent Soviet diesel-powered submarine, the 'Kilo', undoubtedly employs diesel-electric drive to power its single shaft. There is therefore no technical reason why this arrangement should not have been trialled in the 'Tango'.

The weapons carried by the 'Tango' almost certainly correspond with those carried by the 'Victor II'. Besides

conventional pattern-running anti-ship torpedoes and electrically powered anti-submarine homing torpedoes, the 'Tango' is probably equipped to fire the SS-N-15 anti-submarine missile. Given the large size of the submarine and the configuration of the hull forward, weapons stowage must be close to that of its nuclear-powered counterpart.

A standard outfit of mast-mounted sensors is carried. Early units had Stop Light ESM in place of the Brick Pulp/Brick Spit antenna carried by contemporary nuclear boats, but this was later superseded by a new antenna designated Squid Head, which appears to be a cross between the two. It is also interesting to note that the traditional Quad Loop DF antenna was retained in favour of the more recent Park Lamp. Since the latter is generally credited by Western Intelligence with VLF reception, this is indicative of the different mode of operation of the 'Tango', which would spend more time on or close to the surface than its nuclear counterparts of the 'Victor' class. Similarly there is no sign of hull doors in the after deck casing which might conceal a VLF buoy.

Eighteen units of the 'Tango' class were completed at the Gorky shipyard from 1972 to 1982, at an approximate rate of two per year. Most serve with the Soviet Northern Fleet, although they generally undergo refit at Kronshtadt in the Baltic. A handful serve in the Black Sea.

Above: *Most units of the 'Tango' class serve with the Soviet Northern Fleet, and would be tasked with protecting the SSBN bastion in the Barents Sea. (MoD, 1987)*

Right: *The 'Charlie II' has a longer bow section than the 'Charlie I', and the hatches covering the launch tubes have a different configuration. Four of the six torpedo tubes are visible above the surfaced waterline in this photograph. (MoD)*

The 'Charlie II'

In 1973 a new variant of the 'Charlie' emerged. As with the 'Victor II', the first of which had been completed one year earlier, the 'Charlie II' featured a longer bow section. Overall length was 102 metres, as compared with only 94 metres for the 'Charlie I'. Other dimensions were unchanged, but submerged displacement increased from 4,800 tons to 5,400 tons. The increase in size was almost certainly responsible for a reduction in maximum speed to about 23 knots, given a propulsion plant of fixed output.

In spite of the enlarged bow section, no attempt was made to increase the number of missile launch tubes, which remained at eight. Although of slightly different configuration from those of the 'Charlie I', the hatches that covered the launch tubes were of approximately the same size. Reference sources all agree, however, that the launch tubes beneath them house a different missile, the SS-N-9, which is reported to have undergone initial trials and testing aboard the experimental 'Papa'.

The SS-N-9 (NATO code-name: Siren) is a somewhat larger missile than the SS-N-7, and has almost double the range (60 nautical miles as compared with 35nm). It would therefore have required longer launch tubes than were fitted in the 'Charlie I'. Since the depth of the respective submarine hulls is identical, the increase in length was presumably accommodated by adjusting the angle of the launch tubes. This would partially account for the additional hull section interposed between the after end of the missile hatches and the fin. It seems logical to assume that the increase in

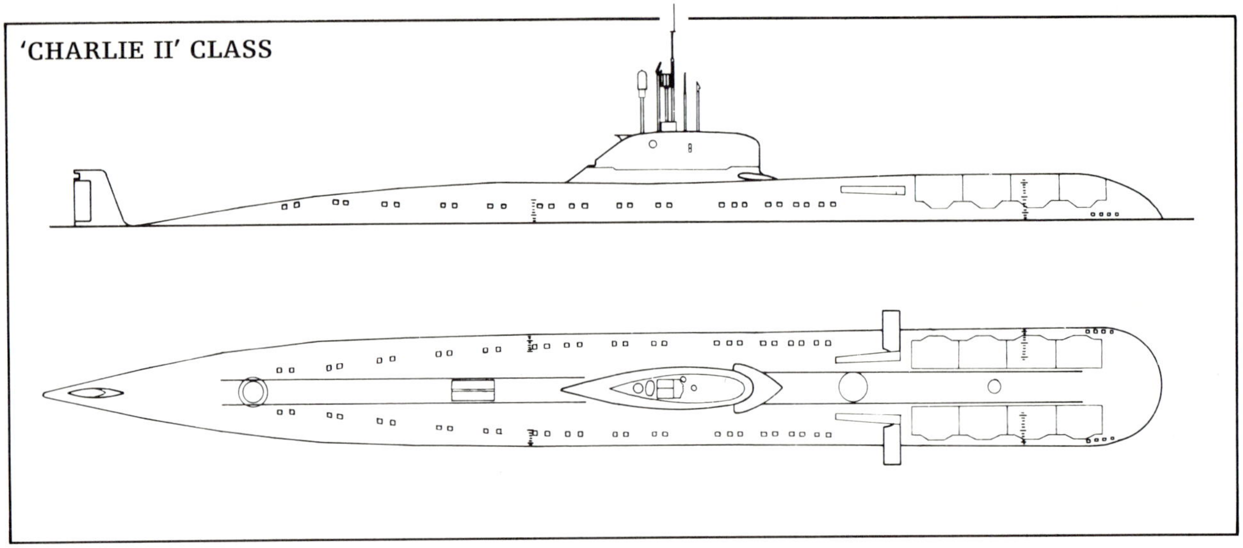

Left: *Note the twin doors for a VLF communications buoy, and the different configuration of the missile hatches. The hydrodynamic 'collar' at the base of the conning tower has been fitted in both variants of the 'Charlie'.*

Right: *Close-up of the conning tower and bow section of a 'Charlie II'. Note the configuration of the missile hatches and the position of the forward hydroplanes. A number of Clusterguard anechoic hull tiles appear to be missing.*

internal volume is otherwise accounted for by additional electronics associated with fire control for the SS-N-9 missile and, as in the 'Victor II', for the SS-N-15 anti-submarine missile.

The mast-mounted sensors fitted in the 'Charlie II' are identical with those of the earlier boats. The only visible modification to the sensor outfit is the provision of twin doors in the after deck casing for a VLF buoy. This again parallels the incremental improvements made in the 'Victor II'. The ability to receive VLF messages at depth would clearly be of considerable tactical value to submarines of this class, as it would enable them to remain concealed in open ocean areas dominated by enemy surface forces until such time as targeting information were available for transmission from the central command post ashore. The submarine would then employ its own sensors to re-acquire the contact.

The 'Charlie II' was a natural development of the 'Charlie I'. The only feature that has puzzled Western analysts has been the comparatively low construction rate – six units over a period of ten years, as compared with twelve units in little more than six years for the 'Charlie I'.

Part of the explanation undoubtedly lies with the change in Soviet submarine building priorities which took place in the early 1970s, when the bastion concept was formulated. The urgent need for submarines designed either to operate within, or to defend the bastions appears to have resulted in a downgrading of the anti-carrier mission. Moreover, construction of the only completely new submarine type of the period, the 'Tango', was allocated to the same Gorky Shipyard that was responsible for the 'Charlie'. These large, sophisticated diesel boats were quickly being turned out at the rate of two per year, and the priority accorded to construction of this class appears to have resulted in a corresponding decline in production of the 'Charlie'.

The other factor that may have been responsible for the tailing-off of 'Charlie' production was the develop-

Left: *Only six units of the 'Charlie II' variant have been completed over a period of ten years. There is every indication that the concept has been overtaken by other developments.*

'CHARLIE II' CLASS

Soviet designation: Podvodnaya Lodka Atomnaya Raketnaya Krylataya (PLARK) (Cruise Missile Submarine, Nuclear-Powered)

Construction:
Built at Krasnaya Sormovo Shipyard, Gorky.
Six units completed 1973–82: one in 1973, one in 1974, one in 1977, one in 1979, one in 1980, one in 1982.

Displacement:
Surfaced: 4,500 tons
Submerged: 5,400 tons

Dimensions:
Length: 102m (335ft)
Beam: 10m (33ft)
Draught: 8m (26ft)

Propulsion:
One pressurized water-cooled reactor driving geared steam turbines; one shaft with 5-bladed propeller; 22,000shp for 23 knots max.

Complement 85

Weapons:
Eight SS-N-9 Siren anti-ship missiles in fixed angle launch tubes; six bow tubes of 533mm diameter for anti-ship/anti-submarine torpedoes and SS-N-15 Starfish ASW missiles; maximum capacity fourteen torpedoes/missiles.

Sensors:
LF active/passive sonar; search and attack periscopes; Snoop Tray surface surveillance radar; Brick Pulp/Brick Spit ESM; Park Lamp DF loop.

Communications:
HF and VHF radio; VLF buoy.

ment of potentially effective countermeasures by the US Navy. By the mid-1970s US Carrier air wings were being organized to include a squadron of ten fixed-wing S-3 Viking ASW patrol aircraft equipped with sonobuoys, sophisticated processing equipment and homing torpedoes, and capable of prosecuting submarine contacts within a radius of more than 100 nautical miles around the carrier battle group. During the same period, the first of a new class of large, quiet, high-performance submarines designed for 'direct support' of the battle groups was completed. The production rate of the *Los Angeles* class would average 3–4 boats per year over the next decade, making it possible to provide every deployed carrier battle group with two submarine escorts capable of clearing a path ahead of the surface ships.

Faced with increasingly formidable anti-submarine defences, the only course of action open to the Soviet Navy was to develop anti-ship missiles which could be fired from underwater at much greater ranges. Greater range implied a major increase in the size of the carrier submarine. The result was the giant 'Oscar'.

The 'Victor III'

The 'Victor III', which first appeared in 1978, was the logical outcome of the bastion strategy. Although developed from a submarine originally intended for the anti-SSBN mission, it incorporated a number of important new features designed to provide a major capability against Western attack boats.

Arguably the most important of these new features was the Soviet Navy's first towed array, dispensed from a distinctive pod-shaped housing located atop the vertical rudder fin. The array, comprising a linear string of passive hydrophones towed at a distance behind the submarine and linked to an on-board processor, would make it possible to detect hostile submarines at much greater range than previously.

Similar arrays already in service in Western navies were already producing remarkable results. The superiority of towed arrays over hull-mounted passive sonars was in part related to the length of the array, which made it easier to establish range and bearing, but was also due to the improvement in the quality of the signal made possible by isolating the array from the machinery and flow noise of the submarine itself.

The principal limitations of a towed array were that it could only be deployed safely in deep, open waters, and that the towing submarine would need to operate at relatively low speed in order for the array to be fully effective. Violent manoeuvres while deploying an array were out of the question, as the array might become tangled and would then have to be abandoned.

The towed array, while it would be useful for tracking ballistic missile submarines, was equally well-suited to patrolling operations in defence of the bastion perimeters in the Northern Fleet and the Pacific Fleet areas.

The other major improvement concerned the weapons capability of the 'Victor III'. The 'Victor II' had introduced the tube-launched SS-N-15 nuclear-tipped anti-submarine missile to the Soviet Navy. The 'Victor III' was designed from the outset to fire its conventional successor, the SS-N-16.

The US Navy had proposed a torpedo-carrying version of its own SUBROC missile as early as 1962, but the missile was never developed. The Soviet SS-N-16 was similar in concept to the US proposal, and is thought to have entered service in 1980. Ironically, it was at about this time that the US Navy renewed its interest in the concept, proposing the Stand-Off Weapon (SOW), with alternative homing torpedo and depth-charge payloads, as a means of exploiting improved sonar performance in a non-nuclear tactical context.

The SS-N-16 (NATO code-name: Stallion) has a maximum range estimated at between 35 and 50 nautical miles. On arrival in the target area, the missile releases a 450mm homing torpedo, which then descends into the water by parachute. When the torpedo reaches a predetermined depth it begins its search pattern. The torpedo (designated E45-75A in one source[1]) is reported to have entered service in 1975 and constitutes the payload of both the SS-N-14 surface-launched anti-submarine missile and the submarine-launched SS-N-16. It has a length of 4.6 metres, and carries a 90kg warhead twice the size of that fitted in the Western Mk 46. It has a maximum speed estimated at 38 knots.

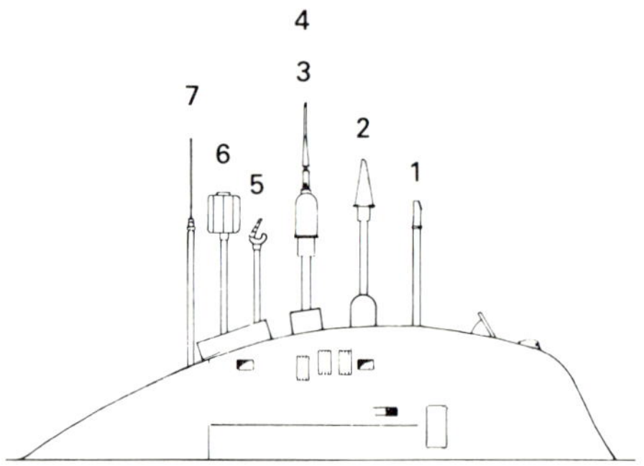

'Victor III'
1. search periscope
2. Pert Spring satellite navigation
3. Brick Group ESM
4. Shot Gun VHF/UHF communications mast
5. Snoop Tray surface surveillance radar
6. Park Lamp VLF reception
7. HF communications mast

'VICTOR III' CLASS

Soviet designation: Podvodnaya Lodka Atomnaya (PLA) (Submarine, Nuclear-Powered)

Construction:
Built at Admiralty Shipyard, Leningrad; Komsomolsk-na-Amur.
23 units completed 1978–88.

Displacement:
Surfaced: 4,800 tons
Submerged: 6,000 tons

Dimensions:
Length: 106m (348ft)
Beam: 10m (33ft)
Draught: 7m (23ft)

Propulsion:
Two pressurized water-cooled reactors driving geared steam turbines; one shaft with co-rotating 4-bladed propellers; two small 2-bladed propellers on stern planes for slow-speed operations; 30,000shp for 29 knots max.

Complement 85

Weapons:
Six bow tubes of 533mm (21in) or 650mm (26in) diameter; anti-ship/anti-submarine torpedoes and SS-N-15 Starfish ASW missiles from 533mm tubes; Type 65 wake-homing torpedoes and SS-N-16 Stallion ASW missiles from 650mm tubes; maximum capacity twenty-four torpedoes/missiles.

Sensors:
LF active/passive sonar; hull-mounted passive arrays; LF towed passive array; search and attack periscopes; Snoop Tray surface surveillance radar; Brick Pulp/Brick Spit ESM; Park Lamp DF loop; Pert Spring satellite navigation.

Communications:
HF and Shot Gun VHF radio; VLF buoy.

Right: *A 'Victor III' running on the surface. The outfit of sensor masts includes the Pert Spring satellite navigation antenna. (MoD)*

Right: *Note the podded housing atop the rudder-fin. In both the later 'Victor' variants the additional hull section has been inserted forward of the conning tower.*

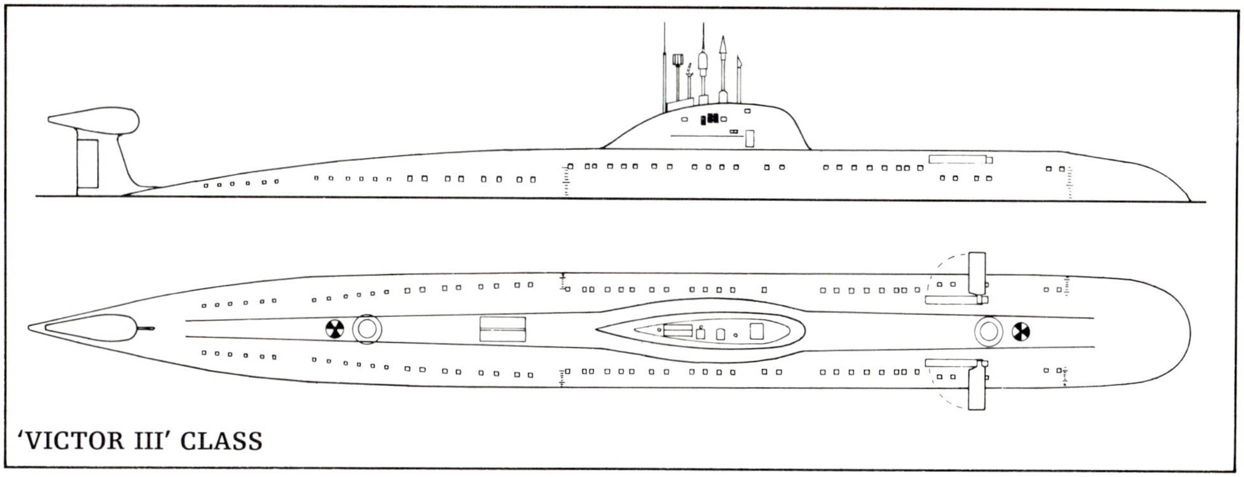

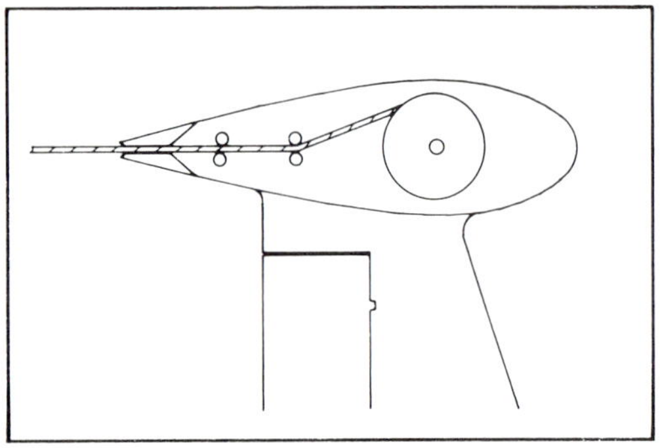

Right: *Close-up of the pod. A number of theories have been put forward regarding its contents, an exotic, magneto-hydrodynamic 'quiet' propulsion system being the most imaginative. Close analysis of the pod has, however, established that it is a dispenser for a towed array, almost certainly for the detection of other submarines. (MoD)*

Left: *Conjectural drawing of the pod, showing the winch for the towed array.*

Left: *The 'Victor III' has a longer bow section than the 'Victor II', and has a different pattern of free-flood holes. The principal distinguishing feature, however, is the tear-drop-shaped pod atop the rudder-fin. (MoD)*

Right: *Stern view of the 'Victor III' which became entangled with a towed array off the coast of North Carolina in November 1983. The trim of the boat has been adjusted in order to free the propeller shaft, revealing the unusual propeller arrangement. Two 4 bladed propellers are bolted on to one another at a relative angle of 22.5 degrees.*

Far left: *The growing interest in anti-submarine warfare displayed by the Soviets during the 1970s is illustrated by the construction of a class of specialized diesel-powered submarines to serve as 'hard targets'. Completed at Komsomolsk in the Far East between 1978 and 1980, the four submarines of the 'Bravo' class are of similar size to the 'Kilo', which followed them on to the slipways. The prominent 'hump' abaft the conning tower is believed to be a padded area designed to receive hits from practice torpedoes. The 'Bravo' is a contemporary both of the 'Victor III' and of the SS-N-16 anti-submarine missile. (MoD, 1986)*

The diameter of the missile itself is thought to be 650mm. It therefore requires a larger-diameter launch tube than the SS-N-15. The 'Victor III' is credited with a mix of 650mm and 533mm tubes (probably four 650mm and two 533mm) in order to accommodate both missiles plus the standard complement of anti-ship and anti-submarine torpedoes.

A new, large-diameter torpedo is reported to have entered service in 1981. Designated the Type 65, the torpedo has an estimated length of about 9 metres, and employs an advanced closed-cycle thermal propulsion system capable of driving it at 50 knots for 27 nautical miles, or at 30 knots for over 50nm. It has an exceptionally large warhead, with alternative conventional or nuclear payloads, and is reported to home on the wake of a ship.

The Type 65 anti-ship torpedo was clearly designed to complement the SS-N-16 anti-submarine missile. Both are fired from the same oversize 650mm torpedo tube, and earlier submarines would therefore have to be specially fitted to accommodate and to fire them. The only submarine type thought to have been so modified is the 'Victor II'. The earlier 'Victor I' probably has inadequate internal volume forward of the fin to accommodate these large weapons plus their associated electronics, although the earlier submarines have almost certainly been retro-fitted with the necessary fire control facilities to enable them to receive the SS-N-15 nuclear missile, which can be fired from a standard 533mm tube.

Further modifications

In November 1983 a unit of the 'Victor III' class ran afoul of a towed sonar array being streamed from a US escort ship off the coast of South Carolina. Subsequently, the submarine had to remain surfaced for several days while other Soviet vessels assisted in salvage operations. The submarine was finally towed to Cuba for repairs.

Attempts to free the propeller shaft, which involved adjusting the trim of the submarine in order to raise the stern out of the water, revealed an unusual propeller arrangement. Twin 4-bladed propellers, angled at 22.5 degrees to each other, were fitted in line on the single shaft, and bolted together. It is reported that not all 'Victor III's are so fitted, earlier units retaining the single 5-bladed propeller of the earlier 'Victor' variants. However, this modification reveals a somewhat crude attempt on the part of the Soviet constructors to achieve a multi-bladed propeller without the benefit of the advanced milling machinery available to the major Western navies. It also serves to explain recent Soviet efforts to obtain such machinery from Western industrial concerns.

Official US Navy statements made before Congress defence committees have indicated that the degree of silencing evident in the 'Victor III' approaches that of the *Sturgeon* (SSN-637) class, which was the standard US attack boat prior to the advanced *Los Angeles* (SSN-688) class. If this assessment is correct, the Soviets must have made a number of incremental improvements in the machinery installation of the 'Victor III' as compared with earlier variants. Indeed, given the exceptionally long production span of the 'Victor' class (some twenty years), it would have been surprising if modifications had *not* been made to the basic propulsion plant. In view of the major changes in submarine operating policy during the period, and in particular the new emphasis on 'quietness', it seems logical that the 'Victor III' would have been used to test some items of equipment and some design features which would be incorporated in future attack submarines.

The only visible modification to the sensor outfit of the 'Victor III' is the addition of a Pert Spring satellite communications mast between the periscopes and the Brick Pulp/Brick Spit ESM antenna. New-model HF and VHF/UHF communications masts were fitted from the outset.

It is of interest to note that whereas construction of both the 'Victor I' and the 'Victor II' was restricted to the Admiralty Shipyard in the Baltic, a second production line for the 'Victor III' was opened up at Komsomolsk during the late 1970s, the first unit being launched in 1978 and completed in 1980. This was made possible by the closure of the 'Delta I' production line, the last boats being completed in 1977. Parallel production of the 'Victor III' at the Admiralty Shipyard and at Komsomolsk was to result in a significant increase in the construction rate. No fewer than 22 submarines of this type were completed from 1978 to 1987, of which almost half were built in the Far East. This provided a considerable boost to the anti-submarine capabilities of the Soviet Pacific Fleet, which until the 1980s had been allocated only a handful of 'Victor I's.

This reorientation of the construction and deployment pattern of advanced Soviet attack submarines is not without significance. At the end of 1981, the first of the US ballistic missile submarines armed with the long-range Trident missile, the *Ohio* (SSBN-726), entered service. Unlike previous US SSBNs, the patrol operations of which had generally been focused on Atlantic waters, the *Ohio* and her sister-submarines would operate out of a new, purpose-built SSBN base at Bangor, Washington, on the west coast of the United States.

Below: *The opening up of a second production line for the 'Victor III' at Komsomolsk has coincided with a major upgrading of the Soviet Pacific Fleet. One of the tasks of the 'Victor III' may be surveillance of the operational bases of Western ballistic missile submarines. (US Navy, 1987)*

The build-up of 'Victor III' construction and deployment in the Far East during the 1980s appears to be closely related to these changes in US SSBN deployment policy. The choice of Komsomolsk as lead yard for the 'Victor III's linear successor, the 'Akula', is particularly significant in this respect. And in 1985 Admiral James Watkins, who was at that time Chief of Naval Operations (CNO) of the US Navy, stated that the frequency of Soviet SSN operations near the Trident base at Bangor 'appears to be increasing',[2] thereby lending further weight to this theory.

It is therefore evident that the Soviets have not abandoned the concept of 'offensive' anti-SSBN operations, and that the 'Victor III' was not intended exclusively for bastion defence. On the contrary, the deployment of virtually all units of the diesel-powered 'Tango' class to the Soviet Northern Fleet area, and the subsequent construction of a new class of diesel boats in the Pacific to perform a similar 'defensive' patrol mission, may indicate an intention to free as many of the nuclear-powered attack submarines as possible for independent, long-range sub-hunting missions. The addition of satellite communication facilities to existing capabilities for the underwater reception of VLF transmissions suggests that such a mission was an important part of the design concept for the 'Victor III'.

7. The Missile Giants

The first unit of the 'Typhoon' class is reported to have been laid down as early as 1975. Built at Shipyard No. 402, Severodvinsk, the hull was launched in September 1980 in an advanced state of completion, and fitting out continued until 1981. The submarine was subsequently involved in a prolonged series of sea trials and missile firings before finally being declared operational in 1983.

The 'Typhoon' made an immediate impact on the Western defence Press because of its immense size. With a submerged displacement estimated at about 26,000 tons, the ballistic missile submarine dwarfs not only its Soviet predecessors of the 'Delta' class, but also the US Navy's *Ohio*, itself a giant in Western terms.

Large size is not generally considered advantageous in submarine operations, which have traditionally favoured stealth and manoeuvrability. A large submarine is easier to detect using active acoustic methods, presents a larger target for anti-submarine weapons, is inherently less manoeuvrable if compelled to take evasive action, and requires deep water in which to operate.

On the other hand, a large submarine possesses greater endurance, has improved habitability for long deployments, and can accommodate a much-increased payload. A large-diameter pressure hull also makes it easier to implement quieting measures, in that potentially noisy machinery can be resiliently mounted and isolated from the hull structure itself.

There are fundamental differences in the respective deployment modes of ballistic missile submarines and attack submarines. Attack submarines deploy against moving targets often concealed in broad expanses of ocean. Large numbers of individual hulls are required in order to perform widely dispersed operations, which may vary from the protection of friendly surface or submarine forces in home waters to the penetration of anti-submarine defences close to enemy bases. Sea-launched ballistic missiles, on the other hand, are deployed against targets of fixed location, and one SSBN with 24 missiles provides the same coverage as two submarines each with twelve missiles. A smaller force of large submarines therefore offers considerable economies in terms of hull construction, logistics, maintenance, and the training of nuclear engineering personnel, without a corresponding loss in effective coverage of potential targets.

Arguments such as these were advanced in the United States to justify construction of the *Ohio* class, which carry 24 missiles (*vice* sixteen in earlier SSBN types) on a displacement approximately twice that of the *Lafayette*s. A similar rationale appears to underly the design of the Soviet 'Typhoon', although in the case of the Soviet boats there were additional factors.

The design of the 'Yankee', which provided the basis for the 'Delta' variants which succeeded it, dates from the early 1960s. During the following two decades, Soviet sea-launched ballistic missiles would grow in size from the 11-metre SS-N-6 to the 14-metre SS-N-18. Even so, missile diameter was constrained by the need to accommodate sixteen launch tubes in a hull that was simply an outgrowth of the 'Yankee'. This limited the number of multiple independently targeted re-entry vehicles (MIRV) that could be carried. Further increases in range and/or payload would be possible only with the development of a completely new missile, the dimensions of which would be free from the constraints of existing hulls and launch tubes.

The SS-N-20 missile (NATO code-name: Sturgeon) constitutes a major advance on earlier Soviet sea-launched ballistic missiles. It is the first production SLBM to employ solid-fuel propulsion, and can deploy from six to nine MIRVs over a distance estimated at 4,500 nautical miles. The CEP (circular error probable) is thought to be half that of the SS-N-18. The missile itself comprises a three-stage rocket with a length of about 15 metres and a diameter of 2.2 metres. It is therefore longer and significantly broader than earlier models.

Development of the SS-N-20 missile was not without problems, probably because of the Soviets' lack of experience with solid-fuel propulsion in missiles of this size. There were numerous failures during testing, and the missile was not accepted into service until 1983, two years after the completion of the first 'Typhoon'.

Hull-form

The 'Typhoon' is essentially designed around the SS-N-20 missile. Twenty launch tubes are located forward of

Right: *The immense beam of the 'Typhoon' is immediately apparent. Overall length is virtually the same as the US Navy's* Ohio, *but the outer hull is almost twice as broad because of the twin-cylinder layout of the inner pressure hulls. Note the massive hydroplanes. (MoD, 1984)*

the fin, and these are completely enclosed within the outer hull casing. The length of the missile determines the depth of the hull, and therefore the overall dimensions of the submarine itself.

Even with double-hull construction, a traditional hull design based on a single cylindrical pressure hull would have implied a diameter of 13–14 metres. The inner pressure hull of the 'Delta' is probably no more than 9–10m, so the figure of 13–14m would have constituted a major increase in dimensions, and would have made the 'Typhoon' inherently less capable of withstanding the pressures of operating at depth. The Soviets therefore adopted the novel solution of incorporating twin pressure hulls side by side within a broad outer hull casing.

The two pressure cylinders each have a diameter of about 9–10m, which is broadly the same as that of the single pressure cylinder of the 'Yankee' and 'Delta'.

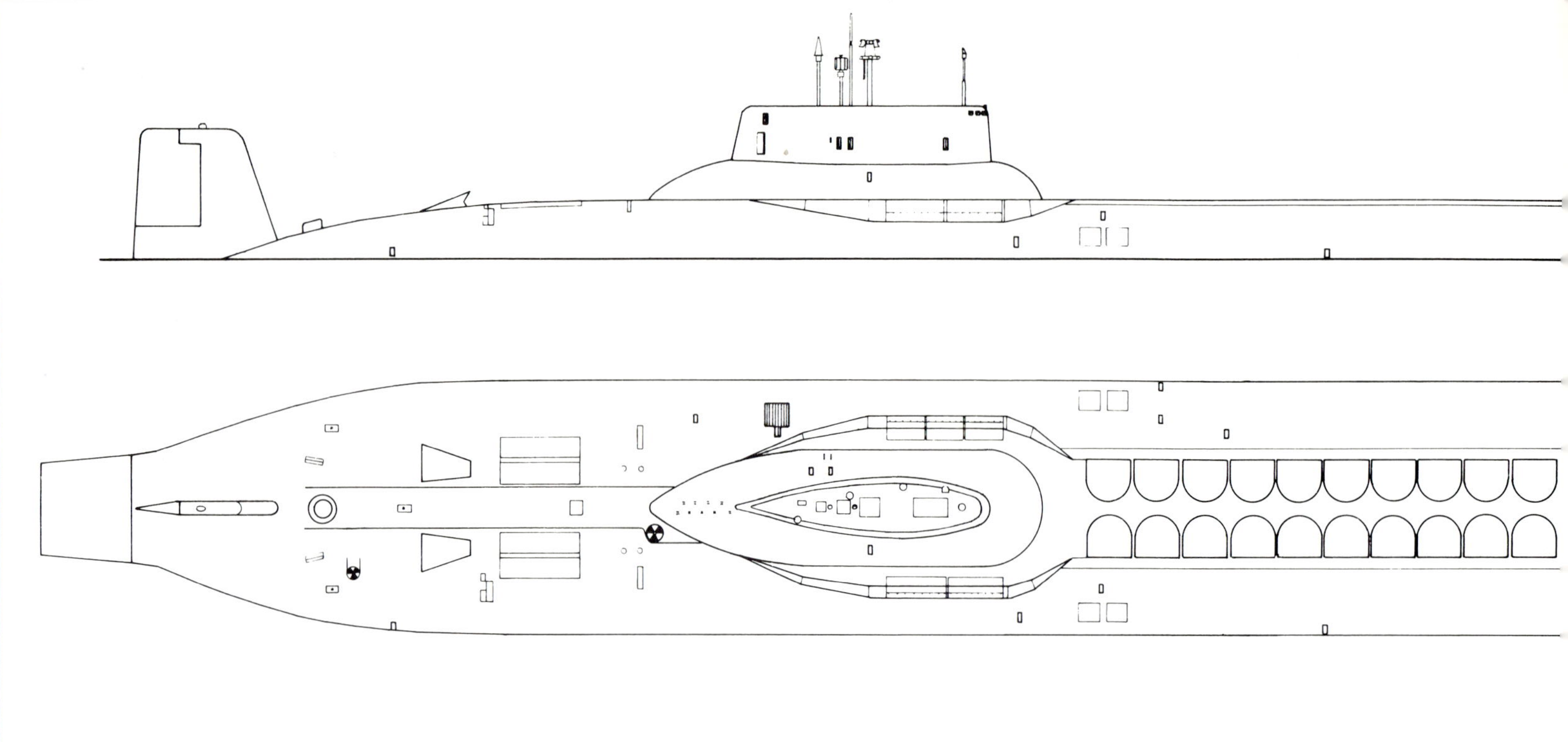

'TYPHOON' CLASS

This would give a similar deep-diving capability, allowing for the same type and thickness of hull steel, and would presumably facilitate the prefabrication of pressure hull sections to standard dimensions. There is thought to be a stand-off of at least 2 metres between the inner pressure cylinders and the outer hull casing, making these submarines inherently more survivable than comparable Western types if struck by a conventional lightweight homing torpedo.

It is the unusual arrangement of the pressure hulls that is responsible for the major differences in dimensions and displacement between the 'Typhoon' and the US *Ohio*. In overall length and draught the two classes are broadly similar. The beam of the 'Typhoon', however, is fully 24m as compared with only 12.8m for the single-hull *Ohio*.

The adoption of twin pressure hulls has made possible a novel layout of the propulsion plant. Each of the two cylinders houses an independent propulsion unit comprising a single pressurized water-cooled nuclear reactor of conventional design, together with a complete set of turbine machinery and its associated gearing. Each propulsion unit powers a single shaft terminating in a 7-bladed propeller. The turbogenerators and other auxiliary machinery are presumably similarly distributed between the two pressure hulls.

This arrangement of the machinery has a number of advantages. It would enable the submarine to continue operations even if a complete propulsion unit had to be closed down as a result of failure or action damage. Moreover, power supply would be relatively unaffected by the closing down of a single propulsion unit, and cross-connections between the two electrical switchboards would enable repairs to be carried out in the cylinder that suffered failure.

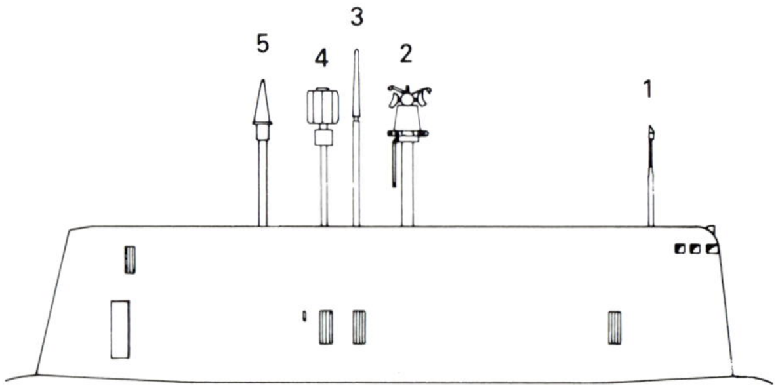

'Typhoon'
1. attack periscope
2. Snoop Pair/Rim Hat surface surveillance/ ESM
3. HF communications mast
4. Park Lamp Mod. VLF reception
5. Pert Spring satellite navigation

Top: *Note the twin sets of VLF hatches set into the after deck casing. The protrusions immediately behind them are reported to house CCTV cameras which monitor the deployment of the VLF buoys.*

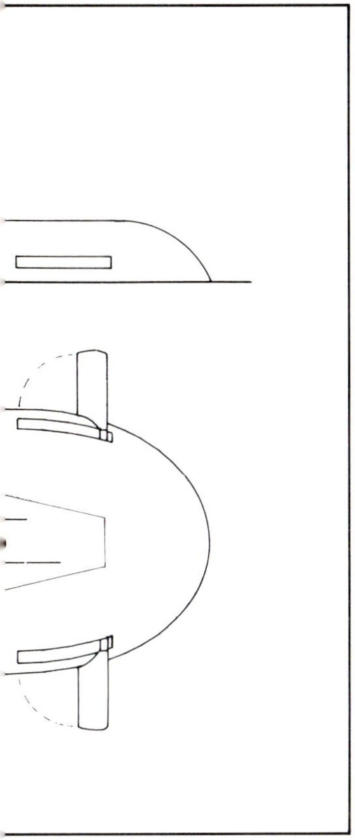

The other major advantage of the machinery layout is the compression of the machinery spaces into a relatively short section of the submarine's hull. Not only can the twin reactors and their related turbine machinery be placed side by side (in all earlier two-reactor submarines they were in line), but additional economies of length may have been effected by dispensing with the emergency diesels installed in the SSBNs of the 'Yankee'/'Delta' series to get them home in the event of reactor failure.

Performance

While the general layout of the propulsion installation is relatively easy to establish, there is less agreement regarding power output with initial assessments ranging from 80,000shp to 120,000shp. These assessments appear to have been based on the assumption that maximum speed is in the region of 30 knots, and the higher figure was presumably influenced by the knowledge that the single reactor which powers the US *Ohio* is rated at 60,000shp.

However, there is nothing in the deployment mode envisaged for the 'Typhoon' to suggest that a speed of 30 knots would be required. The submarine is clearly not designed for high performance; it is simply a floating missile battery, intended to be trundled around in 'safe' waters. High speed was therefore a much less pressing requirement for the submarine's designers than endurance, invisibility and quiet operation.

It is, moreover, open to question as to whether the 'Typhoon' could accommodate the massive 60,000shp reactor of the US *Ohio*, given its unconventional twin pressure-hull configuration. Each of the pressure cylinders is some 3 metres less in diameter than the single pressure hull of the American SSBN, and the relatively short stern section of the 'Typhoon' appears to preclude two smaller reactors in line within each of the cylinders. Maximum reactor size is probably about 35–40,000shp, for a total of 70–80,000shp. The reactor may therefore be based on the model fitted in the icebreakers of the *Arktika* class, which is rated at 28MW (equivalent to 37,500shp).

Because of the unconventional hull configuration of the 'Typhoon', wetted surface area is approximately 65 per cent greater than that of the *Ohio*. The 'Typhoon' has a number of additional features that increase drag; the bow is fully rounded, not elliptical as in Western SSBNs, and the massive fin is at least four times as large as that of *Ohio*.

Both the bow and the fin of the Soviet boat have been streamlined, and there are surprisingly few free-flood holes in the casing for a submarine of double-hull construction. Nevertheless, if all the negative features of the 'Typhoon' hull configuration are taken into account, it becomes clear that even 80,000shp would be insufficient to drive a submarine of this size above 25 knots, and that maximum speed may well be less, perhaps 22–24 knots. It is of interest to note in this respect that the US Navy's *Ohio*, the design of which was based on the same guiding principles of stealth and endurance, is officially credited only with a speed 'in excess of 20 knots'.

Weapons and sensors

The positioning of the SS-N-20 launch tubes forward of the fin is unique in ballistic missile submarine design. Earlier SSBNs of both Soviet and US design placed the missile compartment amidships with the fin well forward.

This arrangement undoubtedly owed something to the fact that the first US SSBNs of the *George Washington* class were conversions of *Skipjack*-class attack submarines already under construction. It made sense to insert the 40-metre missile section at the mid-point of the submarine in order to avoid upsetting the balance of the original design. Placing the missile section at the mid-point also minimized the trim problem which results from firing the missiles, with the consequent ingress of water into the missile tubes.

The *George Washington* conversion proved so suc-

'TYPHOON' CLASS

Soviet designation: Podvodnaya Lodka Atomnaya Raketnaya Ballisticheskaya (PLARB) (Ballistic Missile Submarine, Nuclear-Powered)

Construction:
Built at Shipyard No. 402, Severodvinsk. Five units completed 1983 onwards: one in 1983, one in 1984, one in 1985, one in 1986, one in 1987, three further units building.

Displacement:
Surfaced: 18,500 tons
Submerged: 26,000 tons

Dimensions:
Length: 170m (558ft)
Beam: 24m (79ft)
Draught: 13m (43ft)

Propulsion:
Two pressurized water-cooled reactors driving geared steam turbines; two shafts with 7-bladed propellers; 75,000shp for 24 knots max.

Complement 150

Weapons:
Twenty SS-N-20 Sturgeon ballistic missiles in vertical launch tubes; six–eight bow tubes of 533mm (21in) or 650mm (26in) diameter; anti-ship/anti-submarine torpedoes and SS-N-15 Starfish ASW missiles from 533mm tubes; Type 65 wake-homing torpedoes and SS-N-16 Stallion ASW missiles from 650mm tubes; maximum capacity 30–36 torpedoes/missiles.

Sensors:
LF active/passive sonar; hull-mounted passive arrays; search and attack periscopes; Snoop Pair/Rim Hat surface surveillance radar/ESM; Park Lamp Mod DF loop; electro-optical devices.

Communications:
HF and Shot Gun VHF radio; 2 VLF buoys; ELF towed array.

Left: *Stern quarter view of the 'Typhoon'. Note the distinctive 'hump' at the base of the conning tower, where the pressure cylinder housing the control centre is located. Note also the sturdy construction of the conning tower and the vertical rudder-fin, which are thought to aid ice penetration. (MoD, 1984)*

cessful that the same overall layout was repeated in all successive classes of US Navy SSBN, including the much larger *Ohio*s, and was adopted by the Royal Navy and the French Marine Nationale for their own ballistic missile submarines.

The adoption of twin pressure hulls in the 'Typhoon' clearly compelled the Soviet designers to undertake a radical reappraisal of the overall layout of the submarine. In particular, machinery weight was now concentrated much farther aft than in earlier Soviet SSBN types, because of the side-by-side layout of both sets of reactors and turbines. It was therefore logical to shift the other major item of equipment, the missile battery, to the forward half of the submarine in order to compensate.

The launch tubes, which are disposed in two adjacent rows of ten along the centre-line, are covered by hatches of semi-circular configuration which open outwards. The tubes are located between the twin pressure cylinders, the diameter of which is significantly reduced at this point.

The torpedo tubes and the weapon stowage/handling facilities are thought to be located in a separate, full-diameter pressure cylinder, located centrally within the broad bow section, with access to each of the main cylinders to port and to starboard (*see* conjectural drawing). Most sources credit the 'Typhoon' with six or eight weapons tubes, at least four of which are probably of 650mm diameter to enable the submarine to launch the SS-N-16 anti-submarine missile and the

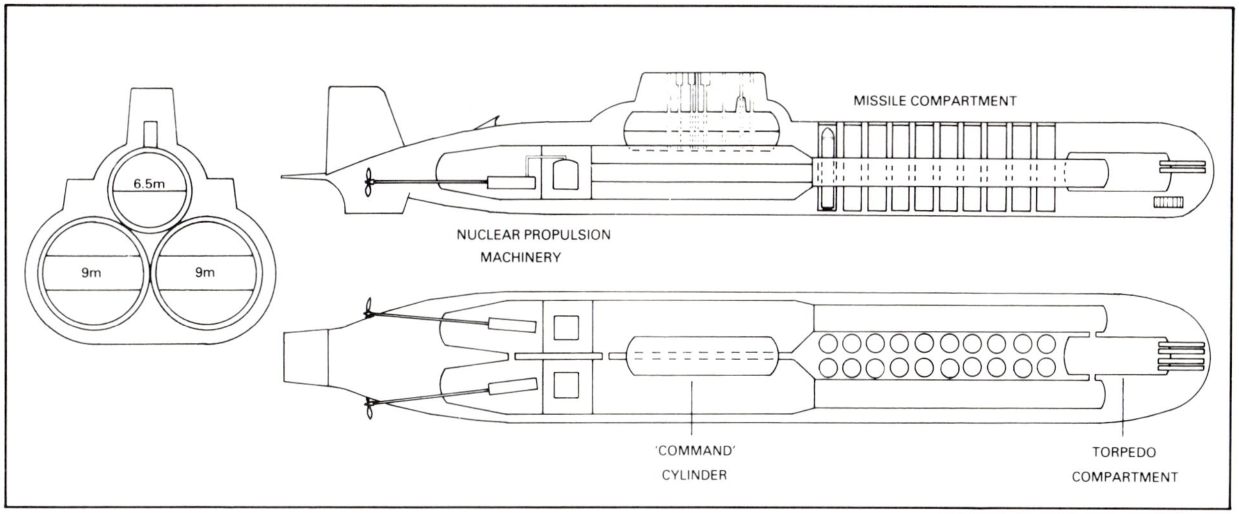

Right: *These conjectural drawings depict the internal layout of the 'Typhoon'. Note the distribution of the nuclear propulsion machinery between the two main cylinders. There is a short cylinder amidships housing the control centre, and a fourth cylinder forward for torpedo stowage and related weapons launch tubes. The vertical launch tubes for the SS-N-20 ballistic missile are located between the main pressure cylinders, which narrow to become little more than passageways at this point.*

Type 65 anti-ship torpedo. The tubes have automatic reloading, and the total weapons capacity is probably about 30–36 torpedoes and missiles.

The bow also houses the large active/passive low-frequency sonar array, probably mounted in the 'chin' of the submarine, and the forward hydroplanes which retract into large apertures just above the surfaced waterline.

Mast-mounted sensors posed a particular problem because of the twin-hull internal layout. Since the sensor masts had to be located on the centre-line they would inevitably retract *between* the twin pressure hulls, making it impossible to use the search and attack periscopes.

The solution adopted was to incorporate a fourth pressure cylinder directly beneath the fin. Not only does the existence of this short, smaller-diameter cylinder resolve the periscope problem, but it explains the large rounded 'hump' in the outer casing at the base of the fin. This 'hump' is penetrated by numerous small frcc flood holcs furthcr cvidcncc of an additional pressure-tight cylinder beneath the fin. It is this fourth cylinder, with a diameter of perhaps 6–6.5m, that houses the submarine's attack centre and communications spaces.

Mast-mounted sensors include search and attack periscopes, HF and VHF/UHF communications aerials, a Park Lamp Mod VLF reception loop, and a Pert Spring satellite navigation cone. The major innovation is the combination of the surface surveillance radar and ESM functions in a bulky mast-mounted antenna designated Snoop Pair/Rim Hat, which has replaced the Snoop Tray and Brick Group antennae of earlier nuclear boats. There are numerous other small hatches both in the top of the fin and in the deck casing, some of which appear to house electro-optical sensors.

Two pairs of twin doors located side by side in the upper part of the after deck casing are thought to conceal VLF buoys, the second buoy presumably being provided for redundancy in the event of the first becoming snagged and lost while deployed (a relatively common occurrence by all accounts). The doors are significantly larger than those fitted in earlier Soviet nuclear submarines, with a length of 7–7.5m and a breadth of 4m. Immediately abaft the doors are prominent angular housings for CCTV cameras, which monitor the deployment of the buoys.

Under-ice operations

From our own perspective, sitting firmly on 55 degrees of latitude, the Arctic Ocean seems a far-off, inhospitable and relatively inaccessible region. From the Kola Peninsula, however, the perspective is somewhat different. The Arctic possesses the dual advantages of physical proximity to the North American continent, and easy accessibility from Soviet Northern Fleet bases, all of which are located within the Arctic Circle. Even more importantly, it offers a vast, and potentially secure haven for friendly missile submarines, and a correspondingly difficult operating environment for Western anti-submarine forces. The area is virtually out of bounds to surface shipping, while ASW patrol aircraft are hampered both by the difficulty of deploying sonobuoys, and by the inability of their homing torpedoes and depth-bombs to penetrate the layer of ice which covers the surface of the waters.

The difficulties which beset surface and air operations in the Arctic region present fewer problems to the nuclear-powered submarine, provided it has the necessary hardening features and is equipped with adequate sensors: downard-looking sonars to establish the

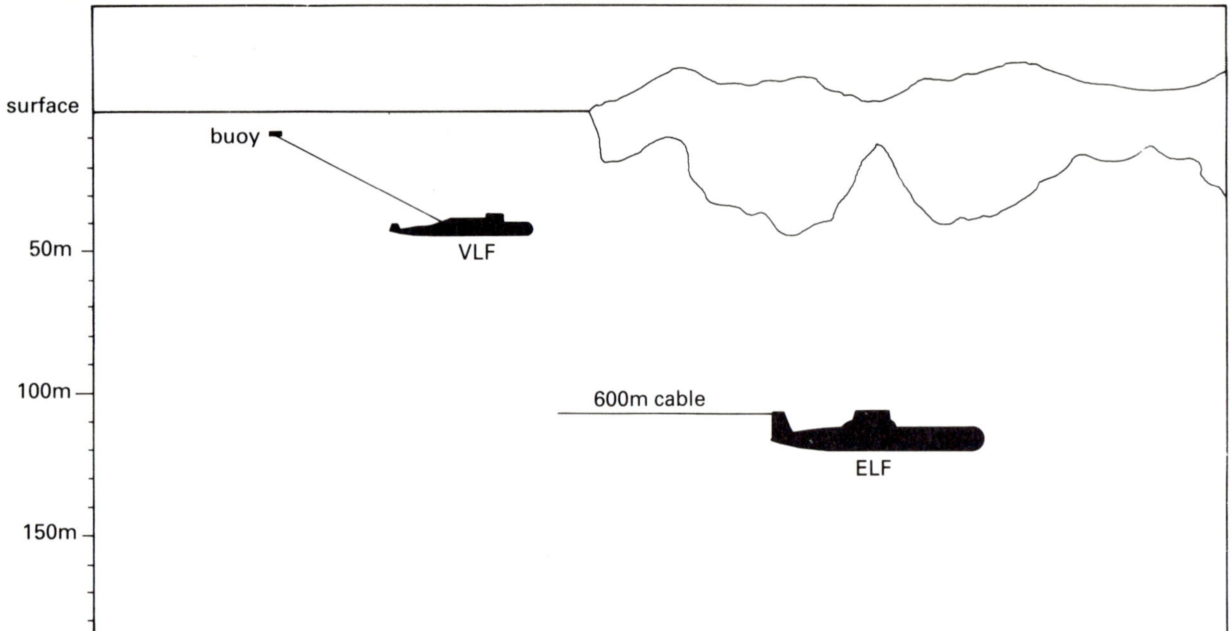

VLF AND ELF: A COMPARISON

depth of water beneath the boat, upward-looking sonars and closed-circuit television capable of determining the profile of the overhead ice, and accurate underwater navigation aids.

The Arctic ice-pack varies considerably in thickness. Even in the coldest months some 3 per cent of the pack is open water. There are also large areas of smooth, relatively thin ice called *polynyas* through which a suitably hardened SSBN could surface in order to launch its missiles.

The 'Typhoon' has a number of design features that make it particularly well-suited to under-ice operations. The long, sturdy fin, which is reinforced at its base by the prominent hump covering the third pressure cylinder, appears to be specifically designed to withstand the impact of surfacing through solid ice prior to missile launch. The prominent tail fin, which is taller and thicker than the vertical control surfaces of earlier Soviet SSBNs, would also assist in ice-penetration manoeuvres. And the forward diving planes, which in the 'Yankee' and 'Delta' classes were mounted high on the fin, have been moved back to the traditional bow position, and are fully rectractable.

The communications outfit of these large submarines almost certainly includes a towed cable for the reception of ELF (Extremely Low Frequency) signals, in addition to the more conventional facilities for VLF reception. The ELF band is particularly well-suited to Arctic operations, as ELF signals penetrate both ice and sea-water to much greater depths than higher-frequency transmissions. In a US Navy test conducted in 1976 using an ELF transmitter in Wisconsin, communication was successfully established with a submarine operating at 16 knots at a depth of 120 metres beneath 9 metres of Arctic ice.[1]

The greater depth at which an ELF cable can be streamed, together with the higher speed at which the submarine can continue to operate, also give the system significant advantages as compared with VLF. A VLF cable has to be streamed relatively close to the surface (15–18m operating depth). The VLF buoy, which is generally preferred by the major navies, enables the submarine to operate at greater depth (45 metres approx.), but has to be towed at a much-reduced speed (generally less than 5 knots), and is liable to snag on ice keels. Under-ice operation also degrades VLF reception to a much greater extent than is the case with ELF signals.

The principal disadvantage of ELF is its slow data rate, which makes it unsuited to delivering missile firing commands or detailed operational instructions. It takes about twenty minutes to transmit a message containing only three elements using ELF, whereas the US Navy's VERDIN system for VLF broadcast transmissions is stated to be capable of 67 words per minute.[2] The other recognized weakness of ELF is that the shore-based transmitters, which require underground cables many kilometres in length, would be vulnerable to attack in wartime, and would be unlikely to survive a nuclear attack, whereas both the Americans and the

Soviets have been able to develop specialized communications aircraft which could continue to make VLF transmissions to their respective ballistic missile submarines after an initial strike.

ELF would therefore be most useful as a 'bell-ringer' during peacetime operations or during the pre-attack phase, advising the submarine to go to shallower depths to monitor transmissions on the LF or VLF frequencies.

It has taken many years of experimentation to develop operational ELF communications systems for submarines. A prototype US transmitter has been conducting evaluation since 1970, and a number of SSBNs have been fitted with experimental receivers. However, serious problems have been experienced in the United States with the siting of the aerials due to environmental considerations and the outright opposition of large sections of the electorate. The Soviet Union, which is blessed with vast expanses of virtually uninhabited territory and a very different political system, has experienced fewer problems in this respect. It is reported that three ELF stations have entered service since 1983, and that these are more powerful than any currently operational in the West.

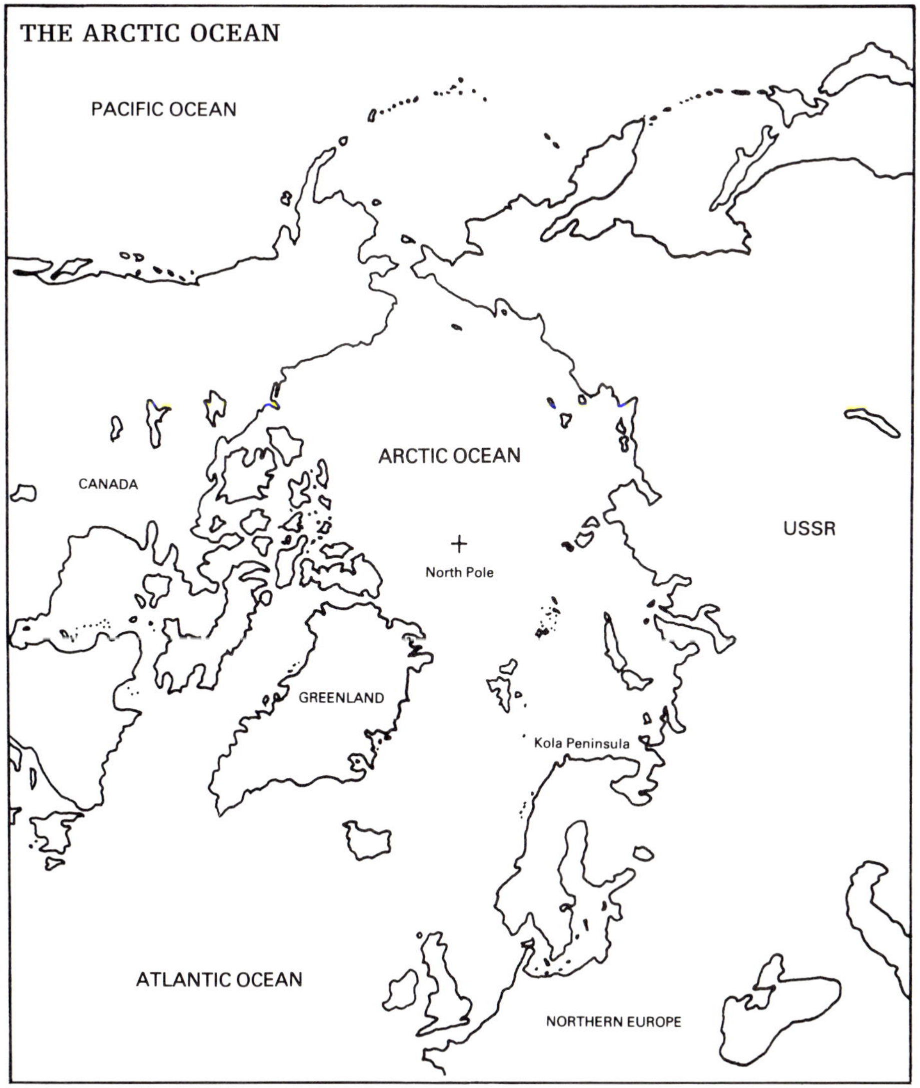

New bases

Two new bases have been identified as being designed to support deployment of the 'Typhoon' in the Arctic region.

The first is Zapadnaya Litsa, located in a deep fiord at the western end of the Kola Peninsula, about 40–50 kilometres from the Norwegian border. Previously thought to be a base for attack submarines, the base at Zapadnaya Litsa has recently been subjected to a major construction programme, and now extends over more than 10 kilometres, with four separate groups of quays and jetties. The harbour area is served by extensive road and rail communications. According to Norwegian Intelligence,[3] Zapadnaya Litsa is regularly used by SSBNs of the 'Typhoon' class.

More significant still is the construction of a major new base at Gremikha, at the eastern end of the Kola Peninsula, 300 kilometres east of Murmansk. Satellite photographs show groups of finger piers each with a length of 200–260 metres projecting from the landward side, and it is reported that docking facilities have been blasted out of the adjacent hillside to provide hardened shelters for submarines.

Gremikha, with its granite cliffs, is well-suited to provide a safe operating haven for the new SSBNs, and is as far to the east of the Kola Peninsula as icing conditions permit. It would therefore be as far removed as possible from the attentions of hostile attack submarines. The hardened docking facilities reveal the degree of concern with which the Soviet Navy regards the possibility of nuclear, or even conventional strikes against its SSBN bases by US submarines armed with the Tomahawk land-attack missile (TLAM). By moving basing facilities as far to the east as possible the Soviets undoubtedly hope to make the task of hostile attack submarines more difficult, in that they would have to pass through Soviet bastion defences in order to take up the most favourable launch positions.

The base at Gremikha, with its granite shelters, appears to be the projected wartime base for the 'Typhoon'. The base at Zapadnaya Litsa, while it may be used by submarines of the class in peacetime conditions, appears far better suited to providing maintenance facilities for the attack submarines

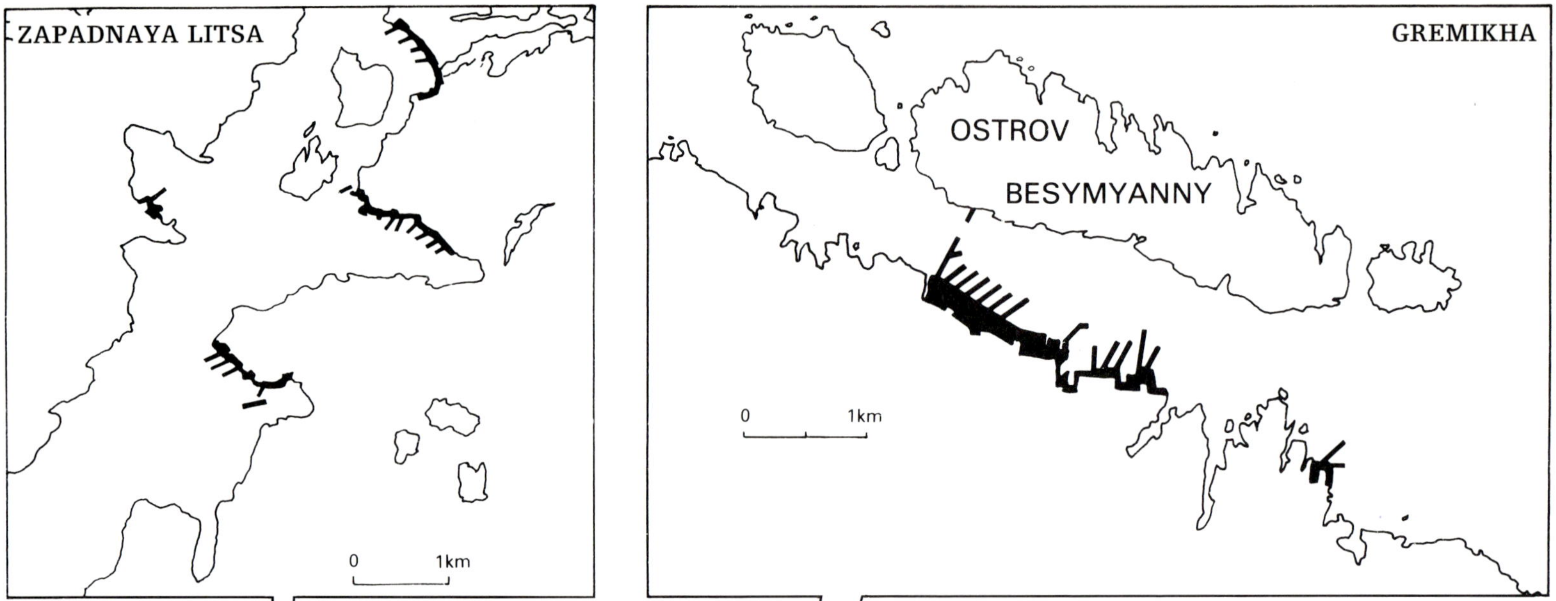

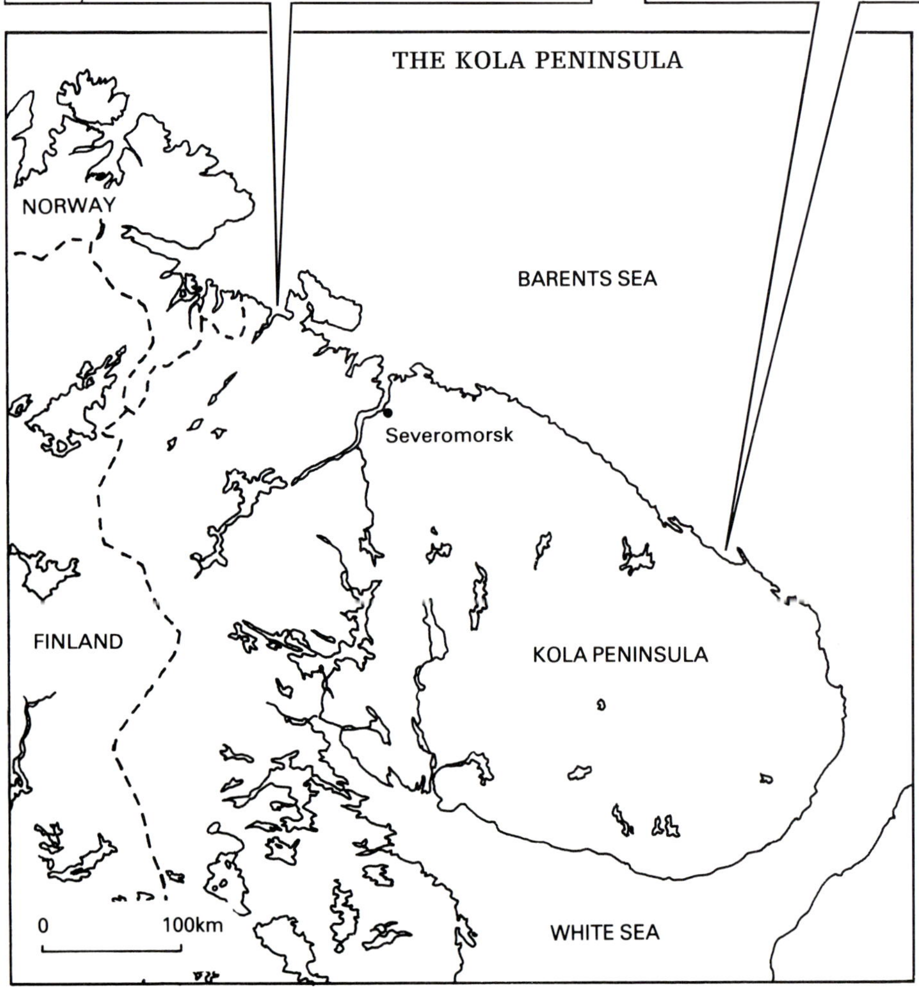

charged either with intercepting hostile surface forces in the Norwegian Sea, or with excluding hostile submarines from the northern bastion.

Either base would provide the 'Typhoon' with easy access to the Arctic ice pack. In the period of protracted conventional operations which many analysts now believe would precede nuclear escalation, the submarines could adopt a purely passive posture once deployed, thereby posing a particularly difficult detection problem to Western attack submarines sent in pursuit. The irregular underside of the ice-pack would enable the SSBN to find suitable protection using a combination of active sonar and on-board closed-circuit television (CCTV). By assuming slightly positive buoyancy it could then come to rest against the underside of the ice, and remain in this position for some time.

Soviet SSBN basing options will be significantly broadened by the deployment of further units of the *Aleksandr Brykin* class, which are capable of transporting and handling a full complement of SS-N-20 missiles. A 75-ton crane enables the *Brykin* to reload the missile tubes of a 'Typhoon' independently of fixed port facilities, thereby opening up the possibility of mobile basing.

Complement

Given the twin-reactor propulsion installation, it is likely that the 'Typhoon' has a crew of at least 150 men. Of greater interest is the reported adoption of the 'Blue/Gold' alternating crew system favoured by Western navies. This provides a significant contrast with crew-

Above: *USN artist's impression of a Soviet submarine base, thought to represent the facilities currently being constructed at Gremikha, at the eastern end of the Kola Peninsula. Hardened shelters have been hacked out of the granite cliffs, and missile submarines of the 'Typhoon', 'Delta' and 'Oscar' classes are moored alongside 'finger' piers 200-250m in length.* (Soviet Military Power)

ing practice for the 'Yankees' and 'Deltas'. The latter submarines appear to have spent a large proportion of their service lives tied up alongside, awaiting the order to move out into bastion areas which would inevitably accompany a major international crisis.

Problems in training an adequate number of highly skilled nuclear engineering personnel have undoubtedly contributed to the adoption of this policy, and there are indications that the Soviets have traditionally had fewer crews than submarines. The change of policy appears to be related to the increased threat posed by US surface and submarine forces to Soviet SSBN bases and to the bastions themselves. The only effective counter to an aggressive US naval power projection policy as embodied in the current Maritime Strategy is to ensure that as high a proportion of the Soviet SSBN force as possible is kept at sea, and that patrol areas are less accessible to anti-submarine operations. Constant patrols beneath the Arctic ice-pack, sustained by a two-crew manning system, are clearly seen by the Soviet Navy as the ultimate solution to the problem.

The 'Delta IV'

The 'Typhoon' is being built exclusively at Shipyard No. 402, Severodvinsk, which is probably the only submarine building facility in the Soviet Union capable of accommodating a hull of this size. At the time of writing five submarines of the class have been completed, and slow series production is continuing. It is thought that the class will eventually comprise at least eight units.

In order to complement these large, sophisticated submarines the Soviets have re-opened the 'Delta' production line at Severodvinsk, completing the first of a new 'Delta IV' variant in late 1984. This occasioned some surprise, as the 'Delta' was by this time a very elderly design, and it was originally thought that the completion of the last of the 'Delta III's in 1982 would mark the end of construction.

There appear to be two motives for re-starting production of the 'Delta'. The first is that the 'Typhoon' undoubtedly has a high initial cost, and construction at Severodvinsk appears to be limited to a maximum of one per year, with each boat taking some five years to complete. The 'Delta', on the other hand, has been in series production since the early 1970s and can be produced relatively quickly and cheaply. Moreover, the 'Typhoon' appears to have been designed specifically

Right: *The new missile transport of the* Aleksandr Brykin *class will make possible mobile basing of the 'Typhoon'. The SS-N-20 missiles are stowed vertically in a large hold forward of the funnel uptakes, and are handled by a 75-ton crane. (RNAF, 1988)*

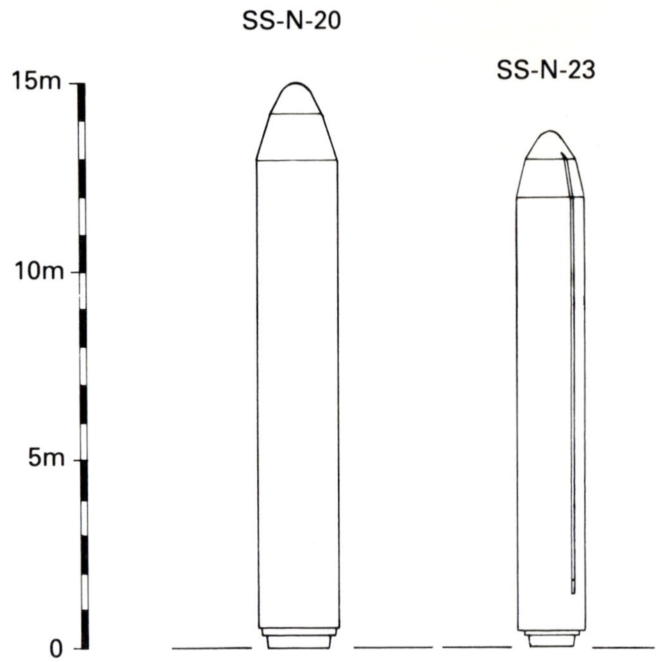

SS-N-20 STURGEON

Length:	15m
Diameter:	2.2m
Propulsion:	3-stage solid-fuel rocket
Guidance:	inertial
Range:	4,500nm (8,300km)
Warhead:	6–9 MIRV × 100KT
In Service:	1983

SS-N-23 SKIFF

Length:	14m
Diameter:	1.8m
Propulsion:	3-stage liquid-fuel rocket
Guidance:	inertial
Range:	4,500nm (8,300km)
Warhead:	7 MIRV × 100KT
In Service:	1986

Right: *Close-up of the tail-fin, showing the towed array dispenser with the rudder beneath. The cable for the towed array runs inside the forward edge of the tail-fin. The winch is almost certainly located between the two hulls, within the prominent angled section abaft the missile casing (MoD)*

for the Arctic theatre, and will not deploy to the Pacific Fleet. The 'Delta' has become the mainstay of Soviet strategic submarine forces in the Far East, and no fewer than seven of the advanced 'Delta III's have been transferred from the Soviet Northern Fleet since the late 1970s. The 'Delta' has therefore become the 'low' end of a 'high/low' mix which is also evident in recent Soviet surface ship construction (e.g., the rocket cruisers of the *Kirov* and *Slava* classes).

The second factor is the development of a new, advanced ballistic missile, the SS-N-23, which constitutes the main armament of the 'Delta IV'. The SS-N-23 (NATO code-name: Skiff) appears to have been designed with a view to restoring the range advantage enjoyed by the SS-N-8 over the SS-N-18, while at the same time retaining the multi warhead capability of the latter missile. It has an estimated range of 4,500 nautical miles (as compared with 3,500nm for the SS-N-18), and is reported to be capable of deploying no fewer than seven MIRVed warheads. Some reliability problems have apparently been experienced, and although the missile has now entered operational service, work to improve some aspects of its performance is continuing.

The SS-N-23 missile is a three-stage missile with similar overall dimensions to the SS-N-18, and employs the same liquid-fuel propulsion. It could easily be retro-fitted in the 'Delta III' as these submarines become due for overhaul. The 'Delta IV' will therefore fulfil a dual role as an important component of the Soviet strategic deterrent forces in its own right, and as a test vehicle for a missile which will be used to update the fourteen 'Delta III's, which will continue to make up a significant proportion of the Soviet SSBN force until at least the turn of the century.

Further modifications

Besides the SS-N-23 missile, there are numerous minor differences in equipment between the 'Delta III' and the 'Delta IV'. Earlier variants of the 'Delta' were fitted with conventional cruciform tail surfaces. The 'Delta IV', however, has a lower, more angular rudder fin, first observed on the second unit of the 'Oscar'-class SSGN (q.v.). At the top of the rudder fin there is a distinctive tubular housing which dispenses a towed array. The new tail-fin is responsible for an increase in the overall length of the submarine to about 160 metres.

A further modification is that in place of the two pairs of doors for VLF buoys, the 'Delta IV' has a single pair of much larger doors, identical in size to those fitted in the 'Typhoon' and 'Oscar'.

At the forward end of the angled section of the missile casing there is a prominent fairing which is reported to house a TV camera with a full view of the after hull casing. This monitors the deployment of the VLF communications buoy (the 'Typhoon' has a similar arrangement). It also presumably helps to resolve the

Below: *Note the single set of large VLF doors in the after missile casing, and the fixed hydrodynamic 'fins' on the stern section.*

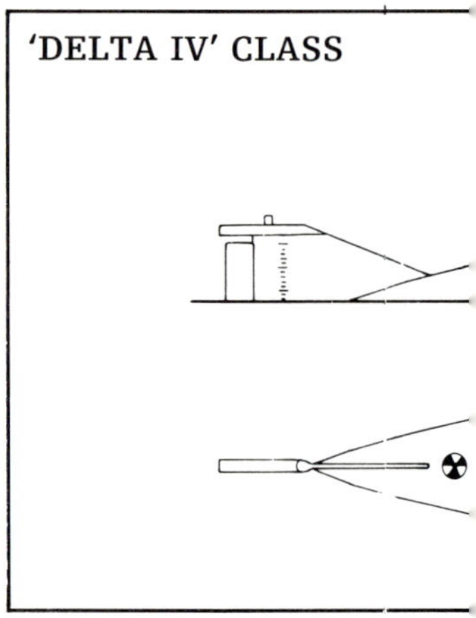

Right: *Early view of a 'Delta IV' in northern waters. Note the CCTV camera housing protruding from the after end of the missile casing, and the remodelled tail-fin. (RNAF, 1985)*

'DELTA IV' CLASS

Soviet designation: Podvodnaya Lodka Atomnaya Raketnaya Ballisticheskaya (PLARB) (Ballistic Missile Submarine, Nuclear-Powered)

Construction:
Built at Shipyard No. 402, Severodvinsk.
Four units completed 1985 onwards: one in 1985, one in 1986, one in 1987, one in 1988; further units building.

Displacement:
Surfaced: 10,750 tons
Submerged: 13,500 tons

Dimensions:
Length: 164m (538ft)
Beam: 12m (39ft)
Draught: 8.7m (29ft)

Propulsion:
Two pressurized water-cooled reactors driving geared steam turbines; two shafts with 7-bladed propellers; 44,000shp for 24 knots max.

Complement 120

Weapons:
Sixteen SS-N-23 Skiff ballistic missiles in vertical launch tubes; six bow tubes of 533mm (21in) or 650mm (26in) diameter; anti-ship/anti-submarine torpedoes or SS-N-15 Starfish ASW missiles from 533mm tubes; Type 65 torpedoes or SS-N-16 Stallion ASW missiles from 650mm tubes; maximum capacity eighteen torpedoes/missiles.

Sensors:
LF active/passive sonar; LF towed passive array; search and attack periscopes; Snoop Tray surface surveillance radar; Brick Pulp/Brick Spit ESM; Park Lamp DF loop; Pert Spring satellite navigation; Cod Eye radio sextant.

Communications:
HF and VHF radio; VLF buoy.

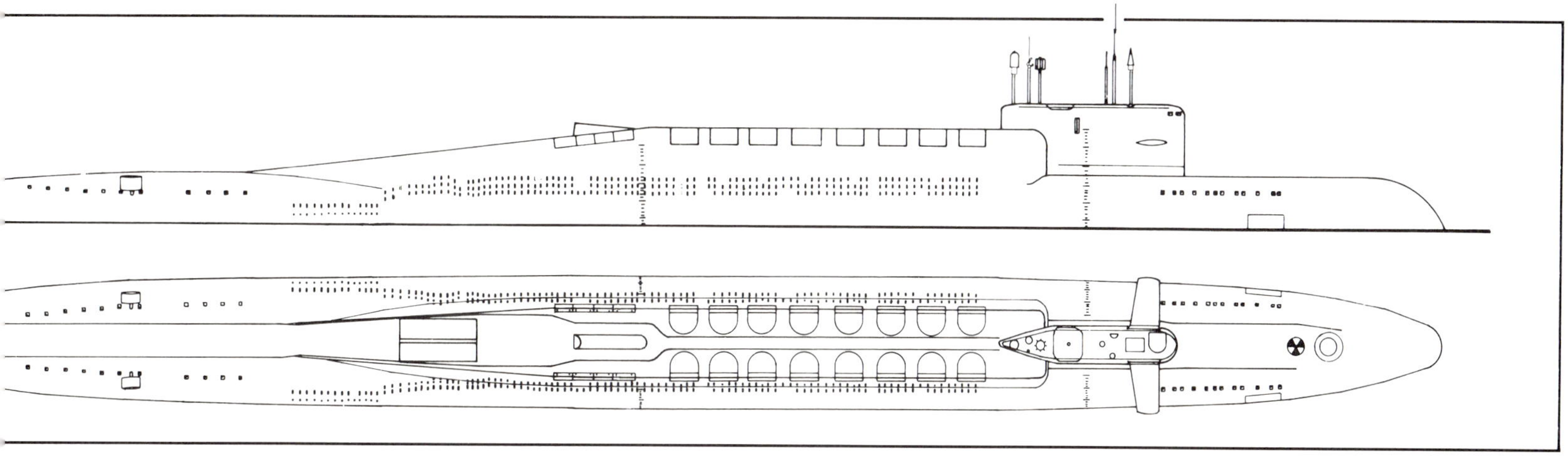

problem noted earlier regarding the restricted view aft from the open bridge atop the fin.

On each side of the after section of the hull there are prominent fixed 'fins' angled upwards at approximately 60 degrees to the horizontal. These appear to have been fitted in an attempt to break up the swirling vortices set up by the prominent missile casing, thereby smoothing the waterflow over the twin propellers.

The 'Oscar'

The last unit of the 'Charlie II' class was completed at Gorky in 1982. At about the same time a new type of nuclear-powered cruise missile submarine (SSGN), the 'Oscar', entered service with the Soviet Navy.

The 'Oscar', however, could in no way be described as a lineal successor of the 'Charlie'. It was a much larger submarine, with a length of 146 metres, a beam of 18 metres (i.e., twice that of the 'Charlie'), and a submerged displacement of about 14,000 tons. In terms of dimensions and displacement the 'Oscar' surpassed the ballistic missile submarines of the 'Delta' class. Such a significant increase in overall size pointed to fundamental changes in Soviet anti-ship tactics as they applied to submarines.

The SS-N-7 and SS-N-9 carried by the 'Charlie I' and 'Charlie II' were relatively small missiles, and consequently had limited range. The submarine would therefore have to approach quite close to a carrier task force in order to launch its missile strike. The advent of the 'Charlie' had a dramatic impact on carrier battle group defences, with a number of new systems being developed to counter both the submarines and their missiles, and the development of new operational tactics (*see* Chapters 4 and 6).

Although there were some American defence analysts who remained sceptical about the effectiveness of these measures, they certainly appear to have convinced the Soviets, who resolved that a new submarine employing new missiles and new tactics would be required if the future effectiveness of the Soviet Navy's anti-carrier forces were to be guaranteed.

US attempts to broaden the diameter of the protective anti-submarine ring around the carrier battle group could be met only by increasing the stand-off range of the anti-ship missile. This was a traditional Soviet response to the problem of carrier battle group defences. It was the rationale behind both the nuclear torpedo and the air-launched missiles deployed on the large maritime bombers of Soviet Naval Aviation.

It was also the rationale behind the SS-N-3 Shaddock missile deployed aboard the rocket cruisers of the

Above: *The large VLF doors and fixed hydrodynamic fins are prominent in this stern quarter view. The CCTV camera housing is not visible, and close analysis of the photograph reveals that it retracts into the missile casing, presumably to improve waterflow over the after end of the missile 'hump'. (MoD, 1988)*

Right: *The sheer size of the 'Oscar' is readily apparent in this bow view of the first unit of the class. The exceptionally broad beam was necessary to accommodate launch tubes for the SS-N-19 anti-ship cruise missile outside the pressure hull. The six square hatches to starboard can be clearly made out abeam the conning tower. Each conceals two launch tubes, angled at approximately 40 degrees. Note also the torpedo reloading hatch above the bow, and the retractable diving planes immediately forward of the missile hatches. Sensor masts in the raised position are, from fore to aft: search periscope, HF radio mast, and Pert Spring satellite navigation antenna. There appears to be a serious problem with anechoic tile adhesion, particularly on the more rounded sections of the hull. (MoD)*

'Kynda' and 'Kresta I' classes and the submarines of the 'Echo' and 'Juliett' classes. The 'Oscar' concept therefore marked a return in some respects to Soviet cruise missile submarine tactics of the 1960s, but there was one major difference. The 200 nautical miles range of the SS-N-3 Shaddock had been necessary because submarines had to launch their missiles from the surface; the SS-N-19 missile which would constitute the main armament of the 'Oscar' could be launched from underwater.

There were numerous advantages in being able to launch at the shorter ranges possible with the SS-N-7 and SS-N-9 missiles carried by the 'Charlie'. Short range meant short flight times. This reduced the 'dead time' between the launch of a missile and its arrival in the target area, and allowed the submarine to re-locate the target using its own passive sensors. It also required a missile body with a relatively small cross-section and a relatively low flight profile, making detection and countermeasures by the target a more difficult proposition. Moreover, a small missile required a smaller submarine to accommodate it, hence the reduction in dimensions between the 116-metre 'Echo' and the 94-metre 'Charlie'.

These advantages were lost with the adoption of the SS-N-19 as the main armament of the 'Oscar'. The 300 nautical mile range of the missile meant a reversion to longer flight times, together with a complete dependence on offboard sensors both for initial detection and targeting of an enemy force. The SS-N-19 (NATO codename: Shipwreck) was significantly larger than the short-range missiles of the SS-N-7/9 series, and therefore required a much larger hull to accommodate it. Moreover, the underwater launch mode meant that the launch tubes themselves would have to be accommodated within the hull, as in the 'Charlie'. The arrangement in the 'Echo' and 'Juliett' whereby the elevating launchers designed for surface launch were stowed horizontally at the outer edge of the deck casing would no longer be possible.

There were, however, significant compensating advantages. A stand-off range of 250–300 nautical miles would place the submarine well outside even the extended anti-submarine defences of the carrier battle group, while underwater launch would preclude a pre-emptive strike by carrier battle group attack aircraft. Moreover, the larger submarine hull required to house the angled launch tubes for the SS-N-19 would be able to accommodate a greater number of missiles. This would make possible a large-scale saturation missile strike, which would be easier to co-ordinate because it could be launched from a single platform. Saturation strikes were now more necessary than ever because of the greater depth and capability of Western anti-missile countermeasures. Furthermore, the increased complement of missiles, together with the larger missile body of the SS-N-19, would make it possible to carry a number of nuclear rounds to complement the missiles with conventional HE warheads.

The adoption of the SS-N-19 missile also established commonality with a new generation of Soviet surface ships designed for the anti-carrier mission. The nuclear-powered rocket cruisers of the *Kirov* class, themselves giant ships, were armed with twenty of the new missiles, launched from angled 'silos' housed within the hull at the forward end of the ship.

Although the 'Charlie' had been dependent on the Soviet ocean surveillance and targeting system for initial detection of the enemy force, its operational mode suggested relative independence from centralized shore control during its run-in on the target, as the presence of anti-submarine aircraft and helicopters would make it a risky business to raise a radio mast above the surface in order to receive detailed position updates and firing instructions, and the submarine's own passive sensors would be adequate for fire control at closer ranges.

The 'Oscar' therefore re-established the traditional Combined Arms approach to anti-carrier operations. It could more easily combine with surface ships and aircraft armed with long-range, stand-off missiles, and was more subject to central control throughout its mission than the 'Charlie'.

Communications

In proposing the Combined Arms approach for forces operating in a maritime environment, the Soviets had long been conscious that the major obstacle to achieving this desirable goal was the difficulty of providing secure, and preferably real-time communications.

The 'Echo' and the 'Juliett' had depended heavily on maritime reconnaissance aircraft such as the Bear D to provide detailed targeting information, and also to serve as a relay for mid-course guidance of the SS-N-3 missile. Both the aircraft and the submarines were subjected to rigid operational control from a central command post ashore. This placed a heavy burden on communications, which had to be conducted using conventional HF and VHF/UHF radio links.

Communications were undoubtedly the weak link in the Soviet ocean surveillance and targeting system. Long-range transmissions on the traditional HF frequency were vulnerable to interception, and short-range 'tactical' transmissions close to an enemy force were subject to jamming. Moreover, communication with submarines was complicated by the need for the submarine to be running on, or just below the surface in order to transmit or receive detailed messages.

The Soviet Navy however has persisted in its belief that effective command, control and communications are fundamental to naval warfare, and has poured massive resources into resolving the communications problem. The development of secure real-time communications would make possible effective centralized control of naval operations, with data from a wide variety of sensor platforms being fed into a computer-assisted command system ashore, and command decisions transmitted almost instantaneously to the various weapons platforms at sea or in the air. Modern technology has enabled the Soviet Navy to take great strides in this direction, and has served to reinforce its belief in effective Combined Arms operations.

The development which perhaps more than any other has made this 'vision' possible is the military satellite. Prototypes were tested from the late 1960s onwards, and by 1974 an operational system was in place. Since that time, the pace of Soviet satellite launches has quickened, and the satellites themselves have become increasingly sophisticated. In December 1987 alone there were no fewer than ten launches of military satellites.

Soviet satellites have a variety of missions, including ocean surveillance, meteorological surveillance, navigation, and tactical communications. Ocean surveillance is generally conducted using a combination of Electronic Intelligence (ELINT) satellites, which pick up the radar and radio emissions of surface ships, and RORSAT (Radar Ocean Reconnaissance) satellites, which employ active radar to investigate contacts established by the ELINT satellites. Some of these satellites now have nuclear power to ensure a longer active life. Photographic reconnaissance satellites capable of producing pictures of increasingly sharp definition have also been developed, although their utility in northern ocean areas subjected to almost constant cloud cover would clearly be limited.

Communications satellites have been used by the Soviets since the mid-1970s, and satcom aerials are now fitted in many classes of surface ship and submarine. Communications via satellite using Ultra or Super High Frequency (UHF/SHF) are relatively secure, because transmissions are 'bounced' off the satellite instead of following the contours of the earth. Also, at the high end of the frequency range, a satellite can transmit data so quickly that the submarine has to spend a minimal amount of time close to the surface.

Satellite navigation was developed by the Soviet Navy at a relatively early stage, as is evidenced by the installation of the Pert Spring satnav aerial in the earliest units of the 'Delta' class. Ballistic missile submarines were clearly accorded priority for the reception of this equipment, but installation of Pert Spring was extended to Soviet attack submarines during the late 1970s, notably to the 'Victor III' subgroup.

The late 1970s also saw the conversion of a number of older boats of the 'Golf' and 'Hotel' classes to communications submarines. The conversion involved removal of the ballistic missile tubes from the fin, and their replacement by additional antennae for tactical VHF/UHF communications and the Pert Spring satellite navigation aerial. In the three modified 'Golfs', a raised structure built as an extension to the after end of the fin houses additional antennae, and there is a prominent fairing flanked by a partially-retractable mast for HF communications at the stern. Further (collapsible) communications aerials can be erected on the deck casing forward. A single unit of the 'Hotel' class has undergone a similar conversion.

The 'Hotel' SSAC serves with the Northern Fleet, as

Below: *One of three units of the 'Golf I' class converted to a communications (SSAC) configuration. Note the extension to the after end of the conning tower, and the prominent elliptical fairing above the stern. (US Navy)*

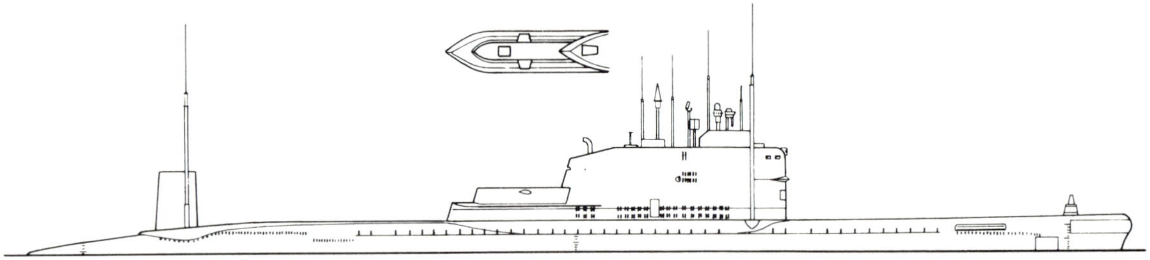

'GOLF SSAC' CLASS

Left: *Most of the new masts are for 'line-of-sight' HF and VHF/UHF communications, but there are new structures immediately abaft the conning tower and close to the stern which house less conventional arrays. There is also a Pert Spring satellite navigation mast.*

Above: *Close-up of the conning tower of a 'Golf' SSAC. Note the Squid Head ESM antenna at the base of the HF radio mast, and the Pert Spring satellite navigation cone. (US Navy, 1986)*

does one of the 'Golfs'; the remaining two 'Golfs' are in the Pacific. All appear to have been modified to serve as command ships for tactical attack submarine operations. Their function would presumably be to serve as a focus for submarine communications within an operational area, and to relay target data and instructions from the central command post ashore. Their ability to submerge would allow them to operate in areas within the operational radius of enemy aircraft – a possibility which would be denied to their surface ship counterparts.

The operational concept that was responsible for the 'Oscar' design took account of all these developments, and a full understanding of their implications is essential in any analysis of the mission and configuration of these large submarines. In many respects the 'Oscar', like its ballistic missile contemporary the 'Typhoon', is essentially a floating missile battery. It is designed not for independent manoeuvre operations on the high seas, in the manner of Western attack boats, but for forward defence to counter US carrier battle group incursions into Soviet sea-space. Operating under rigid central control, the 'Oscar' would take up its designated position in the path or on the flank of a task force, and await the command to launch its missiles. Its extensive communications outfit would enable it to receive constant target data updates via satellite, while the ability to launch its missiles from underwater would confer a degree of security from pre-emptive attack such as would ensure the success of the submarine's mission.

Hull configuration

The tubes for the 24 SS-N-19 anti-ship missiles are located between the outer hull casing and the inner pressure cylinder over a length of about 45 metres – approximately one-third of the overall length of the 'Oscar'. The outer hull is exceptionally broad, with a diameter of 18 metres. The hull casing above the surfaced waterline is also exceptionally flat, suggesting a broadly square hull configuration in the area of the missile compartment (*see* schematic drawing).

The long, low fin is between the two rows of missile hatches. The bow is fully rounded, reflecting the location of the missile tubes in the forward section of the hull. Abaft the missile compartment, however, the hull tapers sharply, and terminates in cruciform after control surfaces.

There are twelve hatches covering the missiles tubes – six on each side of the fin. These hatches are of rectangular configuration, with a length of approximately 7 metres, and are hinged outboard. Two launch tubes are located beneath each hatch for a total of 24

Far right: *Conjectural drawing of the layout of the missile tubes in the 'Oscar'. Note the virtually square cross-section of the hull, and the large stand-off between the pressure hull and the outer casing.*

Near right: *The section of the conning tower casing immediately forward of the Punch Bowl satellite communications antenna houses a cylindrical escape chamber.*

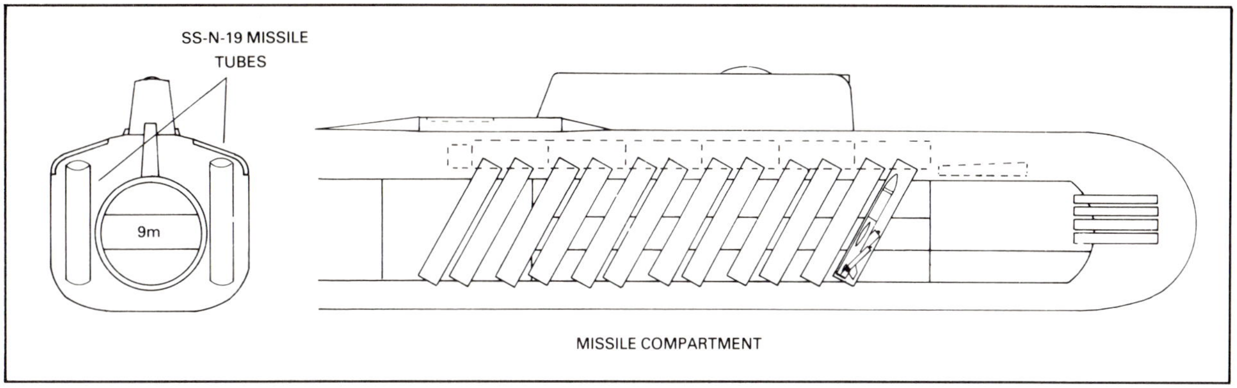

missiles. It is generally agreed that the tubes are not vertical, but are inclined at about 40 degrees. This seems logical, as the SS-N-19 missile has the conventional aeroplane configuration common to other Soviet anti-ship missiles, with solid-fuel boosters slung beneath, and is therefore almost certainly fired from a ramp within the launch tube.

Location of the missile tubes between the outer casing and the pressure hull determines the maximum diameter of the latter, which is probably about 10 metres – approximately the same diameter as the pressure cylinders of the 'Delta' and the 'Typhoon'. Since all three types are built at the Severodvinsk Shipyard, this would facilitate the prefabrication of the pressure hull sections, and would be sufficient for a three-deck layout beneath the fin.

The stand-off between the inner and outer hulls would be about 4 metres in the area of the missile compartment and 2–3 metres abeam the machinery spaces. As with the 'Typhoon', some commentators have suggested that this would make the submarines difficult to disable using standard lightweight anti-submarine homing torpedoes, which have warheads of limited explosive power. But the capability of these large submarines to resist damage has almost certainly been exaggerated, particularly as the SS-N-19 missiles, which are located in a particularly vulnerable position between the two hulls, are reported to have volatile liquid-fuel propulsion.

Sensor and fire control suite

The control and attack centres are located directly beneath the fin, and are therefore conveniently placed between the missile tubes themselves. The outfit of sensor masts is broadly similar to that of other contemporary Soviet nuclear boats, but there are also antennae that are related specifically to the anti-ship mission.

The periscopes, which include search and attack models, are located at the forward end of the conning tower. The radar and communications masts are housed in the after section. These include HF and VHF/UHF communications, a Park Lamp VLF reception loop, and the Pert Spring satellite navigation antenna. Somewhat surprisingly, the distinctive Snoop Pair/Rim Hat combined surface surveillance radar/ESM antenna, which is a feature both of the 'Typhoon' and the latest

Below: *The towed array installed from the second unit of the 'Oscar' class first ran trials aboard one of the elderly 'Novembers'. Note the tubular dispenser set into the vertical stabilizer. (MoD)*

Soviet attack submarines of the 'Sierra' and 'Akula' classes, is not fitted in the 'Oscar', which instead has the smaller Snoop Head/Bald Head antenna previously observed only in the 'Alfa'. Other, smaller hatches of circular or square configuration conceal a variety of antennae, with electro-optical sensors particularly prominent.

In the centre of the fin there is a large semi-circular hatch which conceals a Punch Bowl satellite communications antenna, housed within a distinctive cylindrical radome atop a retractable mast. Punch Bowl can receive targeting data on a direct down-link from Sviet ocean surveillance satellites, and has previously been fitted only in large surface vessels. It is reported that the entire assembly retracts into a pressure-tight housing within the conning tower structure.

Immediately forward of the Punch Bowl antenna there is a prominent bulged fairing. This section of the fin accommodates a spherical escape chamber similar in concept to the Gäbler model fitted in the West German-built SSK-1500 submarines recently completed for India. In the event of the submarine being disabled, members of the crew would enter the chamber via a hatch from the control centre, and the entire structure, including a large section of the outer casing of the conning tower, would lift off from the submarine.

The large twin doors set into the raised section of the hull casing immediately abaft the fin are of similar configuration to those on the 'Typhoon' and the 'Delta IV', and presumably house the same type of VLF communications buoy.

The first unit of the 'Oscar' class was fitted with a conventional vertical rudder, but the second unit was completed with a modified tail-fin which has been a feature of all subsequent units. The new fin is longer, lower, and more angled than that of the first boat, and accommodates a tubular dispenser for a towed array.

After initial speculation that this might be for ELF communications, it now appears that the array in question is identical to that deployed from the tear-drop-shaped pod atop the tail-fin of contemporary Soviet attack boats. The very different configuration of the respective dispensers appears to be simply a matter of the need to accommodate the cable drum for the array within the pod itself in the smaller SSNs. In both the 'Delta IV' and the 'Oscar' there is sufficient space available between the inner pressure cylinder and the outer hull casing to accommodate the cable drum. In the 'Delta IV' the drum may be housed within the 'hump' abaft the missile section, while in the 'Oscar' it appears to be accommodated immediately forward of the tail-fin, abaft the pressure cylinder.

The broad bow section houses either six or eight torpedo tubes and the large active/passive low-frequency sonar. The transducer for the latter appears to be below the tubes, because the upper section of the bow features a prominent set of torpedo reload hatches on the centre-line. Of the six or eight tubes, at least four are presumably of 650mm diameter, and can handle the SS-N-16 anti-submarine missile and the Type 65 wake-

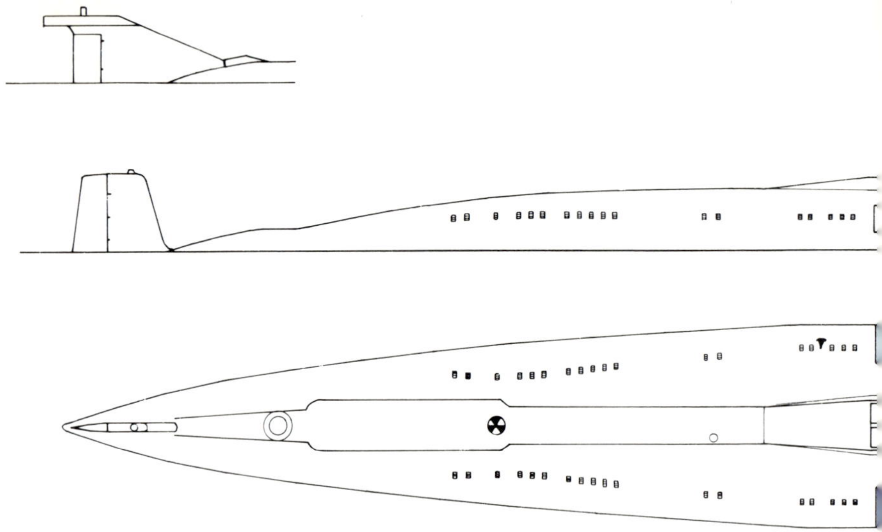

'OSCAR' CLASS

'OSCAR' CLASS

Soviet designation: Podvodnaya Lodka Atomnaya Raketnaya Krylataya (PLARK) (Cruise Missile Submarine, Nuclear-Powered)

Construction:
Built at Shipyard No. 402, Severodvinsk.
Five units completed 1982 onwards; one in 1982, one in 1983, one in 1985, one in 1986; one in 1988 and further units building.

Displacement:
Surfaced: 11,500 tons
Submerged: 14,500 tons

Dimensions:
Length: 146m (479ft)
Beam: 18m (59ft)
Draught: 11m (36ft)

Propulsion:
Two pressurized water-cooled reactors driving geared steam turbines; two shafts with 7-bladed propellers; 75,000shp for 25 knots plus.

Complement 150

Weapons:
Twenty-four SS-N-19 Shipwreck anti-ship missiles in fixed angle launch tubes; eight bow tubes of 533mm (21in) or 650mm (26in) diameter; anti-ship/anti-submarine torpedoes and SS-N-15 Starfish ASW missiles from 533mm tubes; Type 65 wake-homing torpedoes and SS-N-16 Stallion ASW missiles from 650mm tubes; maximum capacity twenty-four torpedoes/missiles.

Sensors:
LF active/passive sonar; LF towed passive array (Hull 2 onwards); search and attack periscopes; Snoop Head/Bald Head surface surveillance radar/ESM; Park Lamp Mod DF loop; electro-optical devices.

Communications:
Punch Bowl SatComm; HF and Shot Gun VHF radio; VLF buoy.

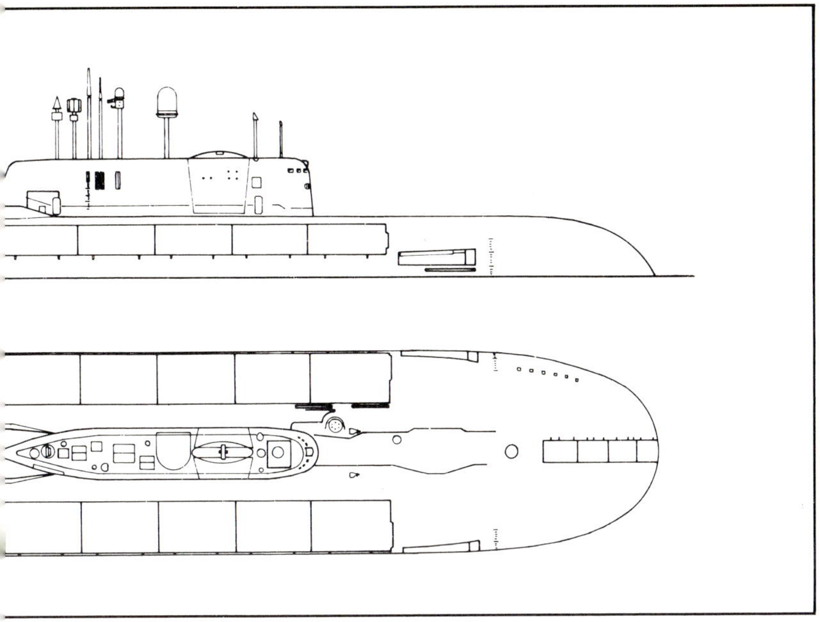

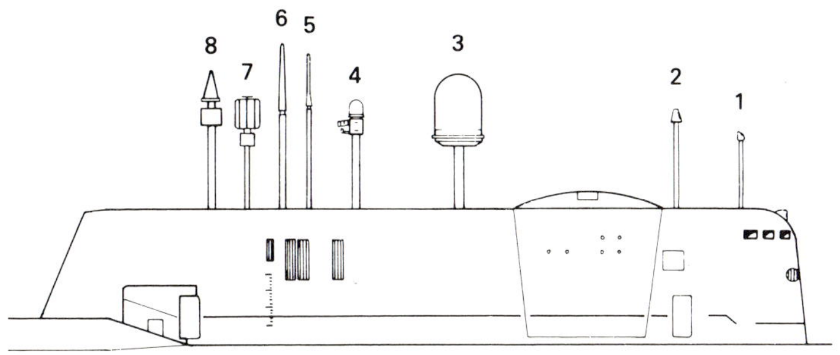

'Oscar'
1. attack periscope
2. search periscope
3. Punch Bowl satellite communications
4. Snoop Head/Bald Head surface surveillance/ ESM
5. Shot Gun VHF/UHF communications mast
6. HF communications mast
7. Park Lamp Mod. VLF reception
8. Pert Spring satellite navigation

Above: *The second unit of the 'Oscar' class, with the modified vertical rudder-fin. Note the bulge in the upper hull casing immediately forward of the tail-fin. (RNAF, 1984)*

homing torpedo. The remaining tubes handle conventional anti-submarine torpedoes and the SS-N-15 nuclear ASW missile.

The large forward hydroplanes are located abeam the torpedo handling room, immediately forward of the missile hatches, and are fully retractable.

Propulsion

Reference sources credit the 'Oscar' variously with one or two nuclear reactors yielding a total of 60,000–90,000shp, for a maximum speed of 30–33 knots. Since the 'Oscar' is a close contemporary of the 'Typhoon', has a pressure hull of similar diameter to the twin

pressure cylinders of the latter, and is built at the same shipyard, it seems logical to assume that the respective propulsion plants are closely related, and may well be based on the same reactor. Unlike other contemporary Soviet attack boats, the 'Oscar' is driven by twin shafts, thereby reinforcing the comparison.

The twin reactors of the 'Typhoon', together with their associated turbine machinery and gearing, are distributed between the twin pressure hulls. The single pressure hull of the 'Oscar', on the other hand, dictates a fundamentally different layout, with the two reactors in line and the turbine machinery aft. This would account for the much greater length of the after section of the 'Oscar' as compared with the 'Typhoon'.

If the propulsion plant installed in the 'Oscar' is indeed closely related to that of the 'Typhoon', a maximum power output of about 75,000shp seems likely. Even this, however, would be inadequate to drive a submarine of the size and configuration of the 'Oscar' at the speeds of over 30 knots suggested by some commentators. Nor does there appear to be any tactical justification for such high speed, given that the submarine would be pre-positioned on the basis of data from off-board sensors in the path of an approaching task force, and would be involved in only minimal manoeuvring. Stealth rather than speed would be required for such a mission in order to prevent detection by hostile fixed-wing anti-submarine patrols. It therefore seems probable that the maximum speed of the 'Oscar' is under 30 knots, and is probably little more than 25 knots.

From the third unit onwards overall length has been increased by some 10 metres, an additional section being inserted abaft the fin. The reason for this modification is not yet clear, although there has been speculation to the effect that it is related to further quieting measures.

Above: *The first four units of the 'Oscar' class have entered service with the Soviet Northern Fleet. Given the age and tactical limitations of the 'Echo II', however, it seems probable that later units of the 'Oscar' class will be allocated to the Pacific Fleet to serve in the anti-carrier role. (MoD, 1987)*

Construction and deployment

At the time of writing, five submarines of the 'Oscar' class are in full commission, with further hulls building. Construction will inevitably be slow, given the large size of the submarines and the requirement for the even larger 'Typhoon' to be built alongside the 'Oscar' at the same Severodvinsk Shipyard.

The five units currently in service are allocated to the Northern Fleet, a deployment which befits their primary mission of defending Soviet territory against seaborne assault. In time of conflict, they would probably be based at the new facilities at Zapadnaya Litsa, close to the Norwegian border, whence they would sortie to intercept US Navy carrier battle groups operating in the Norwegian Sea.

The transfer of the second unit of the *Kirov* class, *Frunze*, to the Far East, however, suggests that as more of these large submarines become available one or two may be transferred to operate against US carrier battle groups in the Pacific.

8. The New Attack Boats

With the completion of the last units of the 'Foxtrot' and 'Juliett' classes during the late 1960s, production of conventional diesel-powered submarines for the Soviet Navy all but ceased. Construction of the 'Tango', which had a specialized mission, was begun at about this time, but the production rate of these large, complex submarines never exceeded two per year. By the late 1970s the Soviet Navy therefore found itself with a large fleet of diesel submarines dating largely from the 1950s and 1960s, and which was fast approaching obsolescence.

Below: *The lead building yard for the 'Kilo' was at Komsomolsk-na-Amur, in the Far East. The submarine was designed as a replacement for older Soviet diesel-powered boats of the 'Whiskey', 'Romeo' and 'Foxtrot' classes. (US Navy, 1986)*

The older diesel boats were still widely used both for training and for local area deployments. The Baltic and Black Sea Fleets continued to depend absolutely on diesel boats to provide a variety of capabilities, including long-range reconnaissance, special operations, and anti-shipping strike in enemy-dominated waters. Submarines with nuclear propulsion were too few in number to perform these tasks, and were too large to operate successfully in these shallow and restricted waters. It was therefore becoming clear that a new conventional design would be required to replace the ageing 'Whiskeys'.

It was not only in the Baltic and Black Sea Fleet areas that such submarines would be required. The Pacific Fleet had for many years operated large numbers of conventional medium submarines. The long-range nuclear boats had generally been based on Petropavlovsk in order to give them free access to the open waters of the Pacific, but the Pacific Fleet area also featured large expanses of enclosed, relatively shallow waters, such as the Sea of Japan, which were equally important to Soviet maritime policy. The growing strength of the Japanese Maritime Self-Defence Force, together with a perceptible shift in the focus of US naval policy away from Europe and towards the north-west Pacific, combined to make the Soviets increasingly apprehensive. The *rapprochement* between the United States and the People's Republic of China served only to reinforce Soviet concerns for the security of the region.

It is perhaps significant to note that from 1978 onwards no fewer than nine submarines of the 'Whiskey' class were transferred from the Baltic and Northern Fleets to the Pacific Fleet via the northern route.

It was therefore not unnatural that the shipyard at Komsomolsk should be chosen as the lead yard for a new type of diesel-powered submarine. The first unit of the 'Kilo' class was duly launched in September 1980, and was completed in April 1982. A second unit was completed the following year, and series production followed.

An 'Albacore' hull-form

The 'Kilo' is remarkable in that it combines many of the features of modern Western submarine construction with specifically Soviet design features.

The hull-form is strikingly similar to that of the US

Barbel class of the late 1950s, and of its lineal successors of the Dutch *Zwaardvis* and *Zeeleeuw* classes and the Royal Navy's *Upholder* class. The overall dimensions of the 'Kilo' are also virtually identical with those of its Western counterparts.

The adoption of the 'Albacore' hull-form appears to mark a final acceptance by the Soviets that submarines, even diesel-powered submarines, are best operated submerged, not on the surface. Surface performance has clearly been sacrificed to obtain superior underwater performance, and the marked reduction in length/beam ratio as compared with earlier Soviet types makes the 'Kilo' more manoeuvrable in restricted waters.

Despite having a submerged displacement 400 tons greater than the 'Foxtrot', the 'Kilo' has a length of only 73 metres – 18m less than the 'Foxtrot', and 4m less than the 'Whiskey'. The other major difference is in beam, which is almost 10 metres, as compared with only 7.5m for the 'Foxtrot' and 6.3m for the 'Whiskey'.

There continue to be major design differences between the Soviet 'Kilo' and contemporary Western submarines. The 'Kilo' is clearly of double-hull construction, and has a prominent flat deck casing. It also has a longer fin than its Western counterparts. The forward hydroplanes retract into recesses in the sides of the deck casing immediately forward of the fin, and not into the space occupied by the main ballast tank forward of the pressure hull, as is generally the case with Western single-hull boats.

Although there is thought to be only a single shaft, terminating in a 6-bladed propeller, the after control surfaces are not cruciform. As with their earlier multi-shaft diesel boats, the Soviets appear to have opted for a conventional vertical rudder beneath the stern, together with horizontal stabilizers incorporating the after hydroplanes.

Double-hull construction must result in some reduction in the diameter of the pressure hull as compared to similar Western boats. A diameter of about 8–8.5 metres seems probable, although the pressure hull appears to be cylindrical throughout most of its length, tapering only at its after end in the region of the motor room. This would make possible a three-deck layout forward and amidships, with the battery compartments on the lower level, and the control centre and accommodation spaces above.

The single centre-line shaft suggests diesel-electric propulsion, with two diesel alternators powering a single large electric motor with a maximum output of about 5,000hp. Battery capacity is almost certainly greater than in earlier Soviet types, enabling the 'Kilo' to sustain high underwater speed for longer periods.

The 'Kilo' is generally credited with six torpedo

'KILO' CLASS

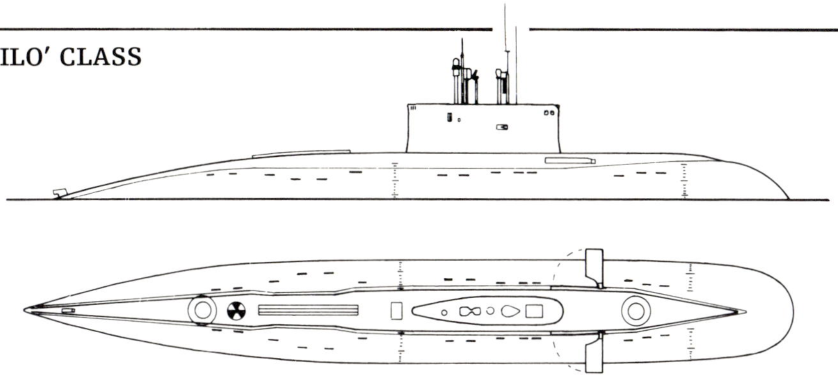

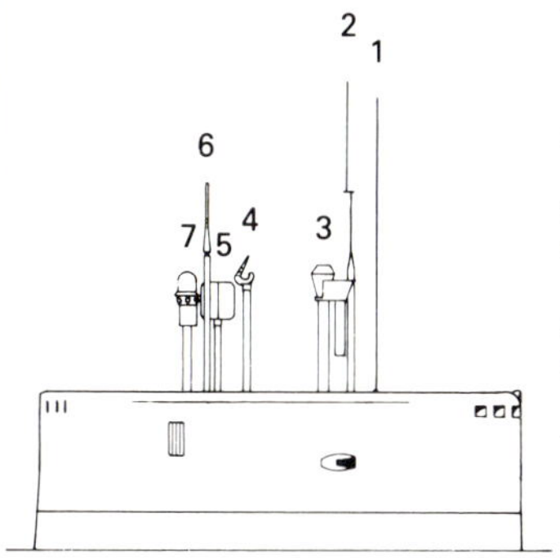

'Kilo'
1. HF communications whip
2. HF communications mast
3. snorkel
4. Snoop Tray surface surveillance radar
5. Quad Loop DF
6. Shot Gun VHF/UHF communications mast
7. Squid Head ESM

'KILO' CLASS

Soviet designation: Podvodnaya Lodka (PL) (Submarine)

Construction:
Built at Komsomolsk-na-Amur; Krasnaya Sormovo Shipyard, Gorky.
Eleven units completed 1982 onwards; 2–3 per year; further units building.

Displacement:
Surfaced: 2,300 tons
Submerged: 2,900 tons

Dimensions:
Length: 73m (240ft)
Beam: 10m (33ft)
Draught: 6.5m (21ft)

Propulsion:
Diesel-electric propulsion; one shaft with 6-bladed propeller; two diesels, 4,000bhp for 12 knots surfaced; one electric motor, 5,000hp for 18 knots submerged.

Complement 54

Weapons:
Six bow tubes of 533mm (21in) diameter for anti-ship/anti-submarine torpedoes; maximum capacity eighteen torpedoes/missiles.

Sensors:
LF active/passive sonar; search and attack periscopes; Snoop Tray surface surveillance radar; Squid Head ESM; Quad Loop DF.

Communications:
HF and Shot Gun VHF radio.

tubes, all of standard 533mm diameter, capable of firing conventional anti-ship torpedoes and electrically powered anti-submarine homing torpedoes. The transducer for the main active/passive sonar is located above the torpedo tubes.

It would theoretically be possible for the 'Kilo' to be fitted to fire the SS-N-15 nuclear anti-submarine missile, but this seems unlikely in view of the submarine's general-purpose mission. Weapons stowage is probably limited to a maximum of twelve reloads, which imposes limitations on the weapons mix. Moreover, the provision of the necessary long-range sonars and fire control electronics would be costly, and would consume valuable space.

The large fin accommodates the standard complement of sensor masts, together with the snorkel, but the antennae are all linear successors to those of earlier diesel boats. There are search and attack periscopes, a Snoop Tray surface surveillance radar, and retractable masts for HF and VHF/UHF communications. More significantly, the Quad Loop DF aerial has been re-

Left: *The 'Kilo' appears to have a rubberized hull coating similar to that of other modern Soviet submarines. Note, however, the different texture of the sheathing around the bow sonar in this overhead view. (MoD)*

Right: *The hull-form of the 'Kilo appears to be modelled on the US Navy's last diesel-powered submarines of the 'Barbel' class and their European successors. Differences include a longer, more substantial conning tower, and the absence of a vertical rudder-fin. Note also the prominent flat deck casing. (US Navy, 1982)*

Above: *The outfit of sensors and weapons is comparatively basic, making the 'Kilo' well-suited to series production, and an attractive proposition for export to friendly countries. The 'Kilo' is now being built in the Western USSR as well as at Komsomolsk, and the production rate has increased from the initial two boats per year to about five per year. (MoD)*

tained in preference to Park Lamp, and the electronic support measures (ESM) antenna is Squid Head, a successor to the early Stop Light model which entered service on later units of the 'Tango' class. There is no evidence of the satellite navigation antennae or the VLF communications equipment fitted as standard in modern Soviet nuclear boats, nor is there any evidence of a towed array capability.

Although the 'Kilo' is of modern design and bears a close physical resemblance to contemporary European patrol submarines, there is evidence to suggest that the equipment, and in particular the electronics, is relatively basic and unsophisticated. This explains the willingness of the Soviets to sell the design in unmodified form to friendly countries such as Poland, Romania, India, Syria and Algeria.

Production lines have now been opened in the western USSR, with submarines being built both at Gorky and at the former Sudomekh Shipyard, now part of the United Admiralty Shipyard at Leningrad. In the light of previous experience, it seems reasonable to assume that Gorky will build units intended for the Soviet Baltic and Black Sea Fleets, while Sudomekh will concentrate on export construction, as it did during the 1970s when it built submarines of the 'Foxtrot' class for a variety of friendly client states.

'Sierra' and 'Akula'

By the late 1970s slow series production of the 'Typhoon'-class SSBN and of the 'Oscar'-class SSGN had been established. The only nuclear-powered submarine type of the 1967–8 generation remaining in full-scale production was the 'Victor', which was by now into its third variant.

The balance of Soviet submarine construction had by this time undergone a dramatic change. Whereas six out of every ten nuclear boats completed throughout the previous decade had been ballistic missile submarines, by 1978 this figure had declined to about two per year. The SSBN programme had delivered the planned number of hulls, and the Strategic Arms Limitation Treaties made further large-scale construction unnecessary. Henceforth it would simply be a matter of replacing older submarines with more advanced types as part of a long-term programme. The emphasis would be on the improved security of the seaborne nuclear deterrent forces, and on more sophis-

Above: *The 'Sierra' was the first of the new generation of Soviet SSNs to enter service. The first hull was launched at Gorky, but was then transferred to the Arctic by inland waterways for fitting out at Severodvinsk. The 'Sierra' has a broader hull-form than the 'Victor III', and has a conning tower markedly different in configuration, but it has a similar pod atop the vertical tail-fin. There are slight differences in shape in the pods fitted in these two classes and the 'Akula'. This has led to suggestions that the configuration may be tailored to the hydrodynamic characteristics of each class. (RNAF, 1984)*

ticated missiles with improved range and throw-weight.

Moreover, the fundamental changes in Soviet SSBN operating concepts had brought with them a requirement for a much larger number of attack boats to enable the Soviet submarine arm to cope with the demands of forward deployment on the one hand, and the defence of the 'bastions' on the other. Fortunately the completion of the major SSBN programme, together with the concentration of 'Typhoon' and 'Oscar' construction at No. 402 Shipyard, Severodvinsk, freed the other Soviet nuclear-capable submarine building yards (notably Komsomolsk) for attack submarine construction. The virtual doubling of the production rate of the 'Victor III', with parallel construction in the Baltic and the Far East, was an early result of this change.

The 'Victor III' was selected in preference to a new type because a rapid expansion of attack submarine production was necessary, and a completely new design would have required a lengthy development programme which would have delayed the submarine's entry into service, and complicated the move to parallel construction in two or more shipyards. Soviet design efforts were therefore initially concentrated on incremental improvements to the 'Victor' which focused on 'quieting', weapons capability, and electronics.

There was clearly a limit to how much new capability could be squeezed into what by now was a relatively elderly submarine. The hull of the 'Victor' had been progessively lengthened to accommodate new systems, but the diameter of the pressure hull had remained the same, and this would preclude the adoption of raft-mounted machinery installation. Improvements in silencing measures would be restricted to the development of better-balanced and quieter-running machinery, together with more effective insulation of the machinery compartments.

It was therefore decided to initiate long-term development work on a new generation of nuclear attack boats which would better meet future requirements and which would incorporate the latest technology.

The first of the new submarines, which received the NATO designation 'Sierra', was duly launched in July 1983 at Gorky and completed the following year. A second new attack boat, the 'Akula', was launched at Komsomolsk in July 1984 and completed in 1985. A third type of nuclear-powered attack submarine, the 'Mike', was launched at Severodvinsk in June 1983, but did not enter service until 1985.

The 'Sierra' and the 'Akula' have already entered slow series production at their respective building yards, and are clearly mainstream types that will constitute the backbone of Soviet attack submarine construction for many years to come.

Both the 'Sierra' and the 'Akula' are larger than the 'Victor III'. Overall length has been slightly increased, but the major difference lies in their beam, which at 12–13 metres is significantly greater than that of their predecessor, and is also responsible for a major increase in displacement.

The parallel development of two distinct types of attack boat of similar size and configuration, with similar weapon and sensor outfits, and in all probability powered by an identical propulsion plant, has puzzled Western observers. It certainly conflicts with the widely held view that uniformity and standardization are the basic tenets of Soviet military equipment provision.

Above: *Although apparently similar in size and mission to the 'Sierra', the 'Akula' has a very different external configuration. The large conning tower is clearly modelled on that of the 'Victor'; it is more rounded and has a long, gently tapered trailing edge. Although the 'Sierra' may have advanced features such as a titanium pressure hull, it is the 'quietness' of the 'Akula' that has aroused most concern in the West.* (Soviet Military Power, *1987)*

A closer examination of Soviet practice, however, suggests that uniformity of design is far less extensive than is generally believed in the West. The myth of Soviet uniformity is certainly contradicted by the plethora of strategic and tactical missiles, all of different shapes and sizes, that are currently in service with the Soviet Navy.

Moreover, the importance and the degree of independence accorded to the various military design bureaux has been underestimated in the West. For many years the Sukhoi and Mikoyan aviation bureaux (OKB) have designed fighter aircraft with similar capabilities for similar roles. In recent years similar developments have become apparent in the design of surface ships for the Soviet Navy. Parallel construction of the 'Kresta II' and 'Kara' BPKs, for service with the Northern Fleet and the Black Sea Fleet/Mediterranean Eskadra respectively, is one notable example. A further example is that of the very different hull-forms and propulsion systems adopted for the destroyers of the *Sovremenny* and *Udaloy* classes, similar in size but designed at the Zhdanov and Kaliningrad Shipyards respectively.

The differences in the external configuration of the 'Sierra' and the 'Akula' may therefore simply be indicative of a desire to strengthen the existing bureaux at the Gorky and Komsomolsk Shipyards by permitting them to produce their own designs within a fairly tight framework of staff requirements, possibly with a view to evaluating the two designs against each other. The introduction of a competitive element into submarine design and construction would undoubtedly bring with it long-term benefits unobtainable with a single-stream programme, as it would be possible to relate specific features of hull design to any deficiencies in performance.

There appears to be a very high degree of commonality as regards equipment between the two types, including the weapons, sensors and propulsion machinery. This reflects the organization of the Soviet military-industrial complex. Standardization exists at the level of production, and the equipment fitted in the two types will undoubtedly be produced at the same inland factories. Where the two types differ is in their hull construction, and in the way the equipment is installed.

Hull-form

Both the 'Sierra' and the 'Akula' have a hull-form broadly similar to that of the 'Victor III', although it appears more squat and compact because of the 2–3 metre increase in beam. Both types also have the 'Victor III' after control surfaces, with the prominent vertical rudder topped by a tear-drop-shaped pod which is believed to dispense a towed linear sonar array.

Here, however, the external similarities end. The 'Sierra' has a low, compact fin with a relatively flat upper section. The 'Akula', on the other hand has a more rounded, streamlined fin with a long, angled trailing edge. Although clearly derived from that of the

'SIERRA' CLASS

Soviet designation: Podvodnaya Lodka Atomnaya (PLA) (Submarine, Nuclear-Powered)

Construction:
Built at Krasnaya Sormovo Shipyard, Gorky.
Two units completed 1984 onwards: one in 1984, one in 1987; further units building.

Displacement:
Surfaced: 6,000 tons
Submerged: 7,550 tons

Dimensions:
Length: 110m (360ft)
Beam: 12m (39ft)
Draught: 7.5m (25ft)

Propulsion:
One pressurized water-cooled reactor driving geared steam turbines; one shaft with 7-bladed propeller; 37,500shp for 30 knots plus.

Complement 90

Weapons:
Six bow tubes of 533mm (21in) or 650mm (26in) diameter; anti-ship/anti-submarine torpedoes, SS-N-15 Starfish ASW missiles and SS-N-21 Sampson land attack missiles from 533mm tubes; Type 65 wake-homing torpedoes, SS-N-16 Stallion ASW missiles from 650mm tubes; maximum capacity twenty-four torpedoes/missiles.

Sensors:
LF active/passive sonar; LF towed passive array; search and attack periscopes; Snoop Pair/Rim Hat surface surveillance radar/ESM; Park Lamp DF loop; Pert Spring satellite navigation.

Communications:
HF and Shot Gun VHF radio.

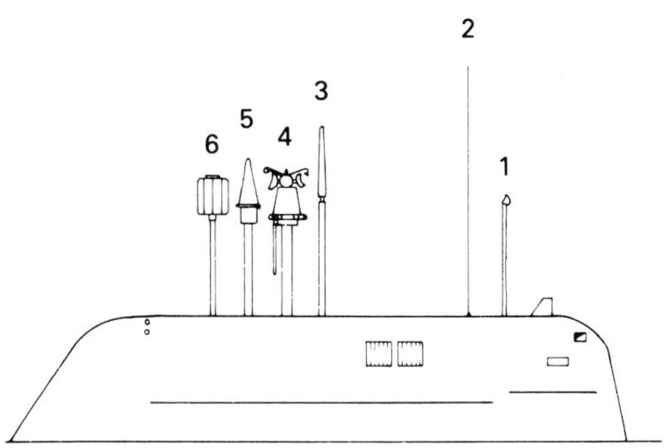

'Sierra'
1. attack periscope
2. HF communications whip
3. HF communications mast
4. Snoop Pair/Rim Hat surface surveillance/ESM
5. Pert Spring satellite navigation
6. Park Lamp VLF reception

Right: *Note the absence of VLF doors in this class, and the position of the forward hydroplanes.*

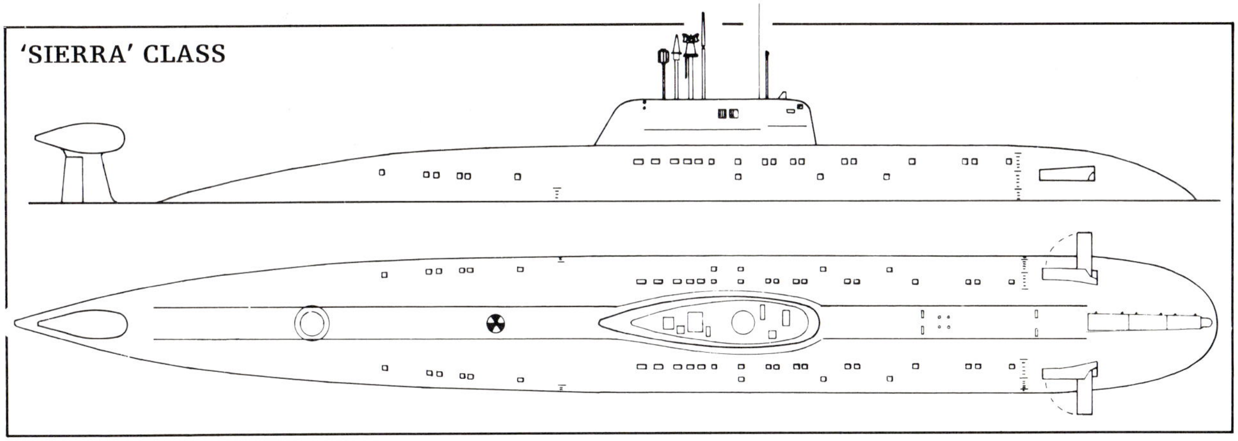

'Victor III', the fin of the 'Akula' has a length of fully 28 metres, as compared with only 19 metres for the earlier boats.

A further important difference is the location of the forward hydroplanes. On the 'Sierra' these are in the conventional position, close to the bow above the surfaced waterline. Those of the 'Akula', however, are not visible in published photographs, suggesting that they are located at the mid-point of the hull, beneath the surfaced waterline, as on the 'Alfa'.

If the 'Akula' is designed primarily for operations in the Pacific area, and the 'Sierra' for the North Atlantic and the Arctic, some of these features may be related to the different operational requirements of those waters. In particular, the fin configuration of the 'Sierra', with its relatively 'hard' upper edges, seems better suited to ice penetration. The Soviets have also generally adopted the upper bow position for the forward hydroplanes of submarines intended for operations in ice-bound waters, but the towed array pod located atop the vertical rudder–fin in both types appears to be vulnerable to damage if surfacing through even relatively thin ice.

Weapons and sensors

Although published photographs have yet to reveal the full array of sensor masts carried by the 'Sierra' and 'Akula', there is every indication that the respective outfits are virtually identical. Besides search and attack periscopes, both types have the bulky Snoop

Above: *An early view of the 'Sierra' in northern waters. Sensor masts, from right to left, are: search periscope, HF radio communications, the distinctive Snoop Pair/Rim Hat combined surface surveillance radar/ESM antenna, and Pert Spring satellite navigation. Note the comparatively short after section, suggesting that the 'Sierra', in contrast with earlier Soviet SSN types, may have only a single large nuclear reactor. (RNAF, 1984)*

Pair/Rim Hat surveillance radar/ESM combination first seen on the 'Typhoon'. This antenna has superseded the separate Snoop Tray surface surveillance and Brick Group ESM antennae fitted in the 'Victor' and other nuclear submarine types of the 1967–8 generation.

There is also a Pert Spring antenna for satellite navigation, a Park Lamp VLF reception loop, and retractable masts for HF and VHF communications. The 'Akula' also has a small drum-shaped ESM antenna (NATO designation: Amber Light) located between the Pert Spring and Park Lamp masts, while the 'Sierra' has a large circular hatch between the periscopes and the after sensor masts which clearly conceals a major antenna, possibly for satellite communications.

The 'Akula' has twin hatches for a VLF buoy in the conventional position atop the after casing, but the hatches are significantly larger than those of the 'Victor III', and therefore presumably house the same model communications buoy as that of the 'Typhoon', the 'Oscar' and the 'Delta IV'.

There is no evidence of similar doors on the after casing of the 'Sierra'. This is a surprising omission, as all Soviet attack submarines completed since 1972 have been fitted with VLF buoys, and the equipment of the 'Sierra' and the 'Akula' appears to be otherwise identical. The absence of the VLF buoy does not, however, preclude VLF communications, as there may be a towed wire antenna, and the mast-mounted Park Lamp aerial is also capable of receiving VLF transmissions, albeit somewhat closer to the surface.

Both the 'Sierra' and the 'Akula' are credited with six weapons tubes. These appear to be located in the conventional bow position, and in the 'Sierra' there is a prominent weapons reloading hatch, similar to that

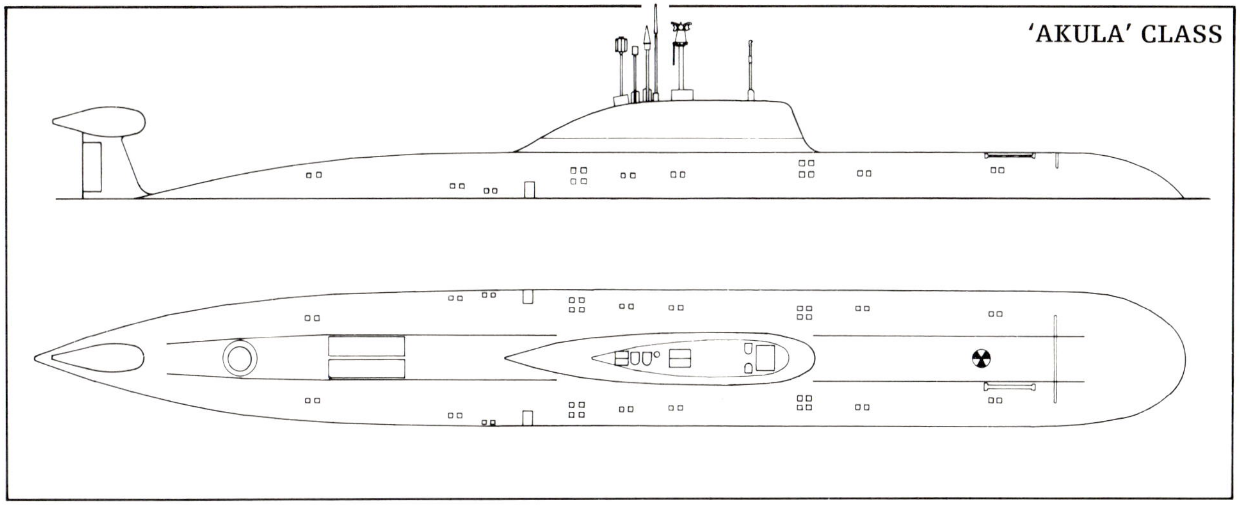

Right: *Note the large VLF doors and the absence of hydroplanes above the surfaced waterline. The conning tower is significantly larger than that of the 'Victor III'.*

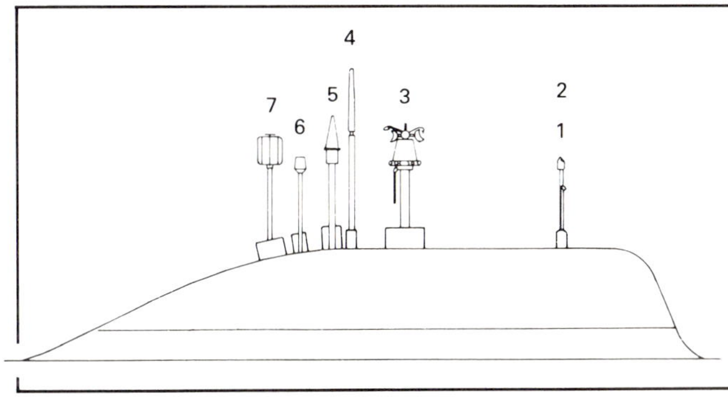

'Akula'
1. attack periscope
2. search periscope
3. Snoop Pair/Rim Hat surface surveillance/ ESM
4. HF communications mast
5. Pert Spring satellite navigation
6. Amber Light ESM
7. Park Lamp VLF reception

on the 'Oscar', running along the centre-line immediately above them. It seems likely that the 'four plus two' arrangement of the 'Victor' has been retained. This would place the LF active/passive sonar transducer (NATO designation: Shark Gill) beneath the tubes, just below the mid-point of the bow.

The SS-N-21

Four of the six weapons tubes are thought to be of 650mm diameter, the remaining two being of standard 533mm diameter. The 533mm tubes can discharge the SS-N-15 nuclear anti-submarine missile and electrically powered acoustic homing torpedoes, while the larger tubes fire the SS-N-16 conventional anti-submarine missile and the Type 65 wake-homing torpedo.

In the early 1980s a new missile, which appears to have been specifically intended for the new generation of Soviet attack submarines, began development. It was tested aboard a specially fitted submarine of the 'Victor III' class, and the missile is thought to have finally entered operational service in early 1988.

The SS-N-21 (NATO designation: Sampson) is similar in concept to the US Navy's Tomahawk Land Attack Missile (T-LAM), which entered service one year earlier. Tomahawk is fired from a conventional torpedo tube in a special container capsule which is jettisoned after launch. It then follows a pre-programmed flight path, using a combination of inertial navigation on the sea leg of its flight, and terrain contour matching, in which the terrain overflown by the missile is compared with a map stored in the flight computer. This system of guidance makes possible very accurate targeting, although the missile itself is more vulnerable to interception and destruction than a missile with a ballistic trajectory.

'AKULA' CLASS

Soviet designation: Podvodnaya Lodka Atomnaya (PLA) (Submarine, Nuclear-Powered)

Construction:
Built at Komsomolsk-na-Amur.
Three units completed 1985 onwards; one in 1985, one in 1987, one in 1988; further units building.

Displacement:
Surfaced: 7,500 tons
Submerged: 9,000 tons

Dimensions:
Length: 113m (371ft)
Beam: 13m (43ft)
Draught: 10m (33ft)

Propulsion:
One pressurized water-cooled reactor driving geared steam turbines; one shaft with 7-bladed propeller; 37,500shp for 30 knots plus.

Complement 90?

Weapons:
Six bow tubes of 533mm (21in) or 650mm (26in) diameter; anti-ship/anti-submarine torpedoes, SS-N-15 Starfish ASW missiles and SS-N-21 Sampson land attack missiles from 533mm tubes; Type 65 wake-homing torpedoes, SS-N-16 Stallion ASW missiles from 650mm tubes; maximum capacity twenty-four torpedoes/missiles.

Sensors:
LF active/passive sonar; LF towed passive array; search and attack periscopes; Snoop Pair/Rim Hat surface surveillance radar/ESM; Amber Light ESM; Park Lamp DF loop; Pert Spring satellite navigation.

Communications:
HF and Shot Gun VHF radio; VLF buoy.

Estimates of the size and performance of the SS-N-21 appear to be based on its US counterpart. The missile is thought to be of similar size to a conventional heavyweight torpedo, flies at about Mach 0.7, and has an effective range of 900–1,100 nautical miles. It can be fired from a standard 533mm tube.

Like the US Tomahawk missile, there are ground-launched and air-launched variants of the missile. There is no evidence to suggest that there is an anti-ship variant, even though such a missile, with a range of 250 nautical miles and a combination of active radar and anti-radiation homing, has been developed for the US Navy in the form of the Tomahawk Anti-Ship Missile (T-ASM).

The absence of any anti-ship equivalent of the tube-launched SS-N-21 appears to confirm that the 'Charlie' concept has been abandoned in favour of larger, specialized anti-ship missile platforms such as the 'Oscar', armed with batteries of long-range missiles each with a warhead large enough to disable a carrier.

Since the SS-N-21 missile cannot be used against surface ships, it is necessary to speculate on the targets against which it might be fired. Use against cities is highly unlikely, as these would be covered by long-range strategic missile systems. The missile would be particularly useful for making strikes against US military installations on or near the coast. A nuclear warhead seems probable, as this would cause maximum damage over a wide area, but a precision guidance system such as the US TERCOM would enable missiles fitted with conventional HE warheads to be fired against precise targets at a lower level of escalation.

The SS-N-21 is almost certainly capable of being retro-fitted to submarines of the 'Victor' class, and within the next few years it seems probable that all Soviet submarines tasked with long-range surveillance duties close to NATO naval basing facilities, particularly those bases which support the patrols of ballistic missile submarines, will have land-attack missiles as part of their armoury.

Propulsion

According to standard reference sources, both the 'Sierra' and the 'Akula' are powered by twin pressurized water-cooled reactors, and their turbine propulsion machinery drives a single shaft with a 7-bladed propeller. A maximum speed in excess of 30 knots seems likely for both classes, given their mission. This would imply an increase in horsepower over the 'Victor', as the new attack boats have two metres more beam and substantially greater submerged displacement.

If the 'Sierra' and 'Akula' are indeed powered by twin reactors in line, this would imply that a new model capable of producing about 20,000shp had been developed specifically for these submarines, but the same propulsive power could be provided by a single reactor of the type employed in the 'Typhoon' and the 'Oscar'.

It would be perfectly feasible for a reactor of this size to be accommodated in the new attack boats, because the pressure hull is of greater diameter than that of the 'Victor'. Indeed, the pressure hull of the 'Sierra' and 'Akula' must be of similar diameter to those of the missile boats, given the multi-hull internal arrangement of the 'Typhoon' and the need to accommodate missile launch tubes on either side of the pressure hull in the 'Oscar'. The relatively short after section of the 'Akula' also appears to lend support to the possibility of a single-reactor installation.

The major innovation of the new attack boats lies in their adoption of many of the quieting features which have been introduced into Western SSNs since the early 1960s. In particular, it seems probable that the increased diameter of the pressure hull has been utilized to reduce noise emissions associated with the machinery. Turbines, gearing and pumps are almost certainly mounted on a 'raft' and isolated from the hull. It is also likely that improvements have been made in propeller design in order to reduce cavitation at higher speeds. It is not known whether the Soviets have introduced natural circulation, but US Navy officials have recently expressed surprise at the degree of quietness achieved by the 'Akula', which represents a major improvement from the noise levels of earlier Soviet attack boats.

There can be little doubt that Soviet efforts to reduce noise levels have been assisted both by espionage and by technology transfer, but the extent to which the 'Sierra' and 'Akula' designs have benefited directly from Western technological input is difficult to ascertain. The sale of advanced Japanese milling machinery by the Norwegian firm of Kongsberg during the early 1980s may take some time to feed into Soviet propeller manufacture. It is by no means clear that the propellers fitted in the latest Soviet attack boats have been produced using such machinery.

Moreover, the Soviets' own experience in the design and construction of submarines should not be underestimated. There has been particularly full and open discussion of submarine quieting measures in unclassified Western sources, and these have frequently been accompanied by quite detailed drawings of the propulsion installations of operational submarines. Once the basic principles of 'rafting' and other noise-reducing measures are understood, it requires only the will to incorporate the necessary features into a design, plus a certain level of technology.

Rather, the lengths to which the Soviets have been prepared to go in order to acquire Western 'know-how' should be seen as an illustration of the high degree of priority now accorded to near-silent operation. This constitutes a major change of direction away from the high-performance, 'brute force' technology of the 1960s epitomized by the 'Alfa'. The new emphasis on stealth, rather than speed, derives from a strategy which increasingly sees enemy submarines as a primary target of the Soviet attack submarine force.

Construction

At the time of writing only two units of the 'Sierra' class have entered service, with a third launched and fitting out. Considering that the first of the class was completed in 1984, this represents a relatively slow rate of construction.

The pattern for construction of the 'Akula' has been somewhat different. The first boat was completed in 1985, but subsequently underwent an unusually long series of sea trials which continued well into 1988. Further units were completed in 1987 and 1988, and the second entered operational service while the first boat was still conducting trials. The rate of construction has therefore been more rapid than for the 'Sierra', but there are clearly features incorporated in the Pacific-built attack boats that are more technologically advanced and have required more extensive testing and evaluation.

The 'Akula' is thought to have an all-steel hull, whereas there have been suggestions that the 'Sierra' may have a pressure hull of titanium. The 'Akula' may have a more advanced electronics outfit, and certainly appears to have the edge on 'quieting'.

The 'Mike'

The third new Soviet nuclear attack submarine type, the 'Mike', remains something of an enigma, despite the considerable publicity which accompanied the loss of the first and only unit of the class in the Barents Sea on 8 April 1989, as the result of a serious electrical fire.

Launched in May 1983 at Severodvinsk, and completed late the following year, the 'Mike' bore little resemblance to either the 'Sierra' or the 'Akula'. Unlike the latter types, it appears not to have entered series

Below: *The 'Mike' running on the surface in northern waters. Note the relatively long after section, and the sturdy tail-fin reminiscent of the 'Typhoon'.*

'MIKE' CLASS

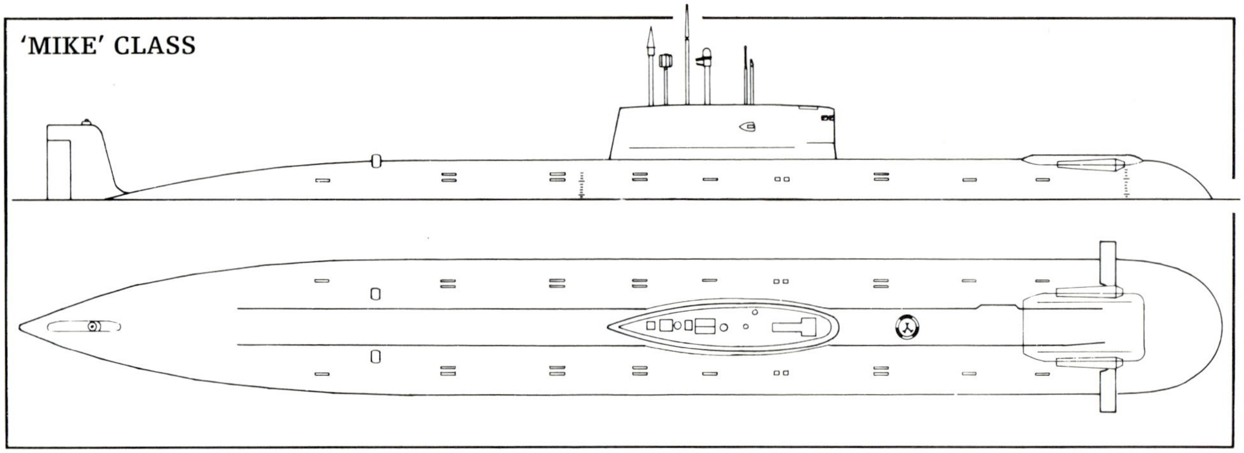

'MIKE' CLASS

Soviet designation: Podvodnaya Lodka Atomnaya (PLA) (Submarine, Nuclear-Powered)

Construction:
Built at Shipyard No. 402, Severodvinsk. One unit completed 1985.

Displacement:
Surfaced: 6,400 tons
Submerged: 8,000 tons

Dimensions:
Length: 110m (360ft)
Beam: 12m (39ft)
Draught: 9m (30ft)

Propulsion:
One liquid-metal reactor with turbo-electric drive; one shaft with 7-bladed propeller; 24,000shp for 25 knots.

Complement 70

Weapons:
Six bow tubes of 533mm (21in) or 650mm (26in) diameter; anti-ship/anti-submarine torpedoes, SS-N-15 Starfish ASW missiles and SS-N-21 Sampson land attack missiles from 533mm tubes; Type 65 wake-homing torpedoes, SS-N-16 Stallion ASW missiles from 650mm tubes; maximum capacity twenty-four torpedoes/missiles.

Sensors:
LF active/passive sonar; search and attack periscopes; Snoop Head/Bald Head surface surveillance radar/ESM; Park Lamp DF loop; Pert Spring satellite navigation.

Communications:
HF and Shot Gun VHF radio.

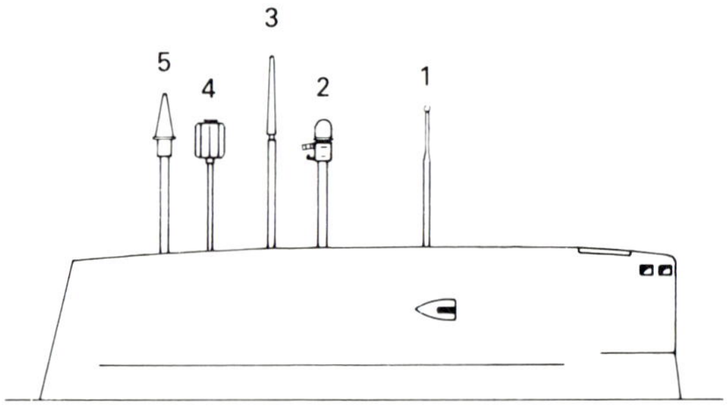

'Mike'
1. attack periscope
2. Snoop Head/Bald Head surface surveillance/ESM
3. HF communications mast
4. Park Lamp VLF reception
5. Pert Spring satellite navigation

production, and is now thought to have been a developmental, 'one-off' design.

Western analysts have tended to see the 'Mike' as the linear successor to the 'Alfa'. This may be true in technological terms, but the difference in size and external configuration is so marked as to suggest that the 'Mike' is a response to a completely new set of operational concepts (as indeed one might expect of a submarine designed some two decades after 'Alfa').

The 'Alfa' was characterized by its very compactness, achieved by combining a lightweight titanium hull with a 'dense' liquid-metal power plant. It remains the only Soviet nuclear-powered submarine type to have a pure 'Albacore' hull-form. The sides of the rounded, fully streamlined fin are faired into the hull casing, and every feature of the submarine is optimized for high underwater speed.

The length of the 'Mike' has been estimated at about 110 metres (initial estimates were even higher) – i.e., an increase of 40 per cent over the 'Alfa'. The hull, unlike the 'airship' hull of the 'Alfa', was a pure cylinder, with a constant beam of 11–12 metres, and had a relatively high length to beam ratio. Moreover the fin, which was both long and low, and in profile closely resembled that of the 'Oscar', joined the deck casing virtually at right angles.

If the 'Mike' continued to employ the titanium construction technology developed at such great expense for the 'Alfa', as has been stated in all open-source publications, it would appear that attention was focused on the strength of titanium rather than its low weight ratio as compared with high-tensile steel, and that the 'Mike' was designed for deep diving rather than high speed. Indeed, some Western commentators have talked of a 'heavy hull', constructed of thick titanium plates.[1]

Certainly the dramatic increase in size, and consequently in hull volume, suggests that the concept of the high-speed 'interceptor' submarine embodied in the 'Alfa' was discarded in favour of a more conventional multi-purpose type capable of conducting patrols of longer duration at greater distances from its home base.

Propulsion

The other revolutionary feature of the 'Alfa' was its advanced liquid-metal reactor, with its high power to weight ratio.

The 'Mike' is also believed to have had a liquid-metal propulsion plant, but it is by no means clear that this was capable of producing the high speeds with which the submarine has generally been credited by standard reference sources. Figures of 60,000shp have been quoted, with a theoretical maximum speed of 38 knots, but it appears that the 'Mike' consistently operated at low speed during her sea trials, and later estimates were of a single liquid-metal reactor of advanced design with a power rating of only 18MW (equivalent to 24,000shp).

It may well be that the liquid-metal reactor of the 'Alfa', whilst it undoubtedly suffered technical problems, showed sufficient promise for the Soviets to persist with development, and that the 'Mike' was engaged in trials of a reactor of new design for possible installation in future Soviet attack boats. The increase in size as compared with the 'Alfa' was probably related to machinery quieting measures. The 'Alfa' has proved to be a particularly noisy boat, and the additional internal volume of the 'Mike', which featured a relatively long after section, will undoubtedly have been utilized to improve this aspect of the submarine's performance.

Below: *The sensor outfit of the 'Mike' is broadly similar to those of the 'Sierra' and the 'Akula'. Seen here, from left to right, are: attack periscope, HF radio mast, Park Lamp DF loop, and Pert Spring satellite navigation. (RNAF, 1985)*

There has been some speculation that the 'Mike' propulsion system may have featured a pump-jet in place of a conventional propeller. Pump-jets for underwater propulsion are already operational in the West, being employed in the US Navy's Mk 48 torpedo, and the latest Royal Navy SSNs of the *Trafalgar* class. The pump-jet features a single multi-blade rotor which turns relatively slowly against stator vanes in a duct. There is some speed loss as compared with a conventional propeller, but this is compensated by the extreme quietness of the pump-jet. It is reported that pump-jet propulsion has been adopted for the new, advanced US Navy SSNs of the *Seawolf* class.

In some respects the 'Mike' may be comparable to the US Navy's *Glenard P. Lipscomb* (SSN-685), which was completed to serve as a propulsion test-bed in 1974. The *Lipscomb*, in which a pressurized water-cooled reactor was allied to turbo-electric drive, was some 20 metres longer than her steam turbine-powered sisters of the *Sturgeon* class. The result was a submarine significantly quieter than the *Sturgeon*, albeit at the cost of some 2–3 knots of speed.

The 'Mike', while employing a more compact liquid-metal reactor, is also reported to have featured turbo-electric drive, which dispenses with conventional steam turbine machinery and its inherently noisy reduction gearing in favour of turbo-generators which supply electrical power to a single large propulsion motor. It is of interest to note that overall length and pressure hull diameter were virtually identical with those of the US Navy boat. The 'Mike' may therefore represent a similar development.

Weapons and sensors

The weapon and sensor outfits of the 'Mike' were similar to those of the 'Sierra' and the 'Akula'. There were six weapons tubes, with a mix of standard 533mm and large-diameter 650mm types, capable of firing the full range of modern Soviet torpedoes and missiles.

There was a prominent hump in the upper deck casing immediately abaft the bow, but the precise purpose of this is not yet clear. The forward hydroplanes, which were mounted unusually far forward, retracted into the lower section of the 'hump', suggesting that the latter may simply have been intended to free bow volume for weapons tubes and the main active-passive sonar.

The full complement of mast-mounted sensors has yet to feature in published photographs, but it appears to have been similar to that of the 'Sierra' with one important exception: the Snoop Pair/Rim Hat surveillance radar/ESM combination was discarded in favour of the smaller Snoop Head/Bald Head array previously fitted in the 'Alfa' and the 'Oscar'.

Notable omissions from the sensor suite were a VLF communications buoy, for which no hatches were provided in the after casing, and the pod-mounted towed sonar array fitted in other contemporary Soviet attack submarines. The vertical rudder-fin of the 'Mike' was entirely conventional, with no tubular wire dispensers.

Designed mission

If the 'Mike' was indeed intended to be the forerunner of a future generation of Soviet attack submarines, it is interesting to speculate on what the primary mission of those submarines might be.

Two main possibilities suggest themselves. The first is that the 'Mike's successor will be designed for long-range operational patrols off the coast of the United States, much as the projected US Navy SSN-21 *Seawolf* is intended primarily for prolonged operations in the Barents Sea. The features of the 'Mike' design that lend support to this theory are its large size, which can be equated with endurance and a substantial weapons reload capacity, its advanced propulsion system, and a capacity for deep diving. The latter features would facilitate penetration of the NATO ASW barriers in the North Atlantic, as the combination of deep diving either with very high speed (implying a twin-reactor plant) or with quiet operation would make the submarine difficult to detect and/or intercept.

The advent of the SS-N-21 land-attack missile makes a long-range mission off the US east coast increasingly attractive to the Soviet military planners. Moreover, the US deployment of submarine-launched Tomahawk nuclear cruise missiles close to Soviet military bases on the Kola Peninsula has provoked clear statements of intent from the Soviet party leadership to the effect that similar systems will be deployed against the United States.

If this is indeed the mission for which the 'Mike's successor is designed, the absence of a VLF buoy communications in the prototype is surprising. If the 'Mike' was intended purely to test the propulsion system for the new submarines, however, this omission is easily explained. There is no evidence to suggest that the 'Mike' ever ventured out of area from its acceptance into full operational service in 1986 to its sinking in April 1989.

The other possibility is that the 'Mike' was the prototype of a new generation of attack submarines intended for Arctic operations. There is a clear requirement for such a submarine, given that a primary mission of the Soviet attack submarine force is to support the deployment of the SSBNs. A number of Western analysts have indeed postulated a scenario in which the 'Typhoons' would deploy under the Arctic ice-pack at the outbreak of a major conflict, and supporting attack boats would be charged with defending the perimeter of the ice-pack against incursions by NATO hunter-killer submarines.

The 'Mike' had a number of features which made the design better suited to such operations than either the 'Sierra' or the 'Akula'. The configuration of the conning tower and of the vertical rudder–fin appear to be well suited to ice penetration. The sturdy vertical rudder–fin, which unlike that of other recent Soviet attack boats, was not fitted with a towed array pod, was in fact remarkably similar in configuration to that of the 'Typhoon' itself.

Below: *The Snoop Head/Bald Head combined surface surveillance radar/ESM antenna is particularly prominent in this view. Note also the 'hump' in the bow casing, and the fixed hydrodynamic fins on the after section of the hull. (RNAF, 1986)*

Postscript

Standing on the threshold of a new era of Soviet submarine construction, it is not easy to predict the course of future developments. If it were simply a matter of projecting recent Soviet technical developments into the future, it would be possible to forecast the nature and balance of Soviet submarine building over the next decade with some degree of confidence.

The accession to power of Mikhail Gorbachev, however, has created shock waves within the Soviet body politic analogous to those that accompanied the brief but stormy administration of Nikita Krushchev. *Perestroika*, and the accompanying shake-up of Soviet institutions, which had become increasingly conservative and complacent under Leonid Brezhnev, will undoubtedly make their mark on the Soviet military-industrial complex, while foreign policy changes will inevitably lead to a fundamental reappraisal of Soviet military strategy.

Below: *Artist's impression of the 'Yankee' converted to conduct trials of the SS-NX-24 land-attack cruise missile. The original missile compartment has been replaced by a 'box' housing twelve launch tubes.* (Soviet Military Power, *1986)*

As in the past, these changes will undoubtedly affect the Soviet Navy far more closely than the other Soviet armed services. Already the deal struck between the Soviets and the Americans on the limitation of intermediate-range nuclear weapons has provoked changes in the deployment of certain types of Soviet submarine in European waters. Even more dramatic are the effects of the US Navy's 'Maritime Strategy' on Soviet Navy deployments and future planning.

Exercises involving Soviet surface ships are currently taking place on a much smaller scale than was previously the case, and are rarely 'out of area'. On the other hand, Soviet submarine activity in the waters around Iceland, and off key US naval bases, appears to have increased. This suggests that, faced with an aggressive US maritime strategy spearheaded by the proposed deployment of multiple carrier battle groups into the Norwegian Sea and beyond, the Soviet Navy has now adopted the policy of conceding the surface of the oceans to the Americans, and grouping its own surface forces in waters dominated by Soviet land-based aircraft, in readiness for a co-ordinated assault against the carrier battle groups when the latter are at their most vulnerable. (A footballing analogy would have the Soviet team conceding the midfield, but packing the goalmouth.)

The main offensive capability of the Soviet Navy would again revert to the submarine arm, which would be virtually the only element of the Soviet Navy operating outside Soviet-dominated waters. The logic of replacing Admiral Gorshkov, whose greatest achievement arguably was to build up the Soviet surface fleet to a level unprecedented in Soviet history, by the submariner Chernavin is therefore irresistible.

Future construction

Soviet submarine construction in the near term can be predicted with reasonable accuracy, if for no other reason than that the types now coming off the building ways have only just entered production. There may, however, be some adjustment of the numbers of each type projected in order to conform to changes in Soviet strategic priorities.

The 'Typhoon' will undoubtedly continue in slow series production, at a rate of perhaps two hulls every three years. It is unlikely that the submarine will be superseded by a new type in the foreseeable future, although there is evidence to suggest that the 'Typhoon' programme will be complemented by continued production of the less complex 'Delta IV', in order to ensure that sufficient new hulls become available to replace existing SSBNs armed with obsolescent missiles.

The 'Oscar' will probably also continue in slow series production for some years yet, although only a limited number will be necessary in order to perform the anti-carrier mission in the North Atlantic. Future production may include units intended for deployment with the Pacific Fleet.

The major new development as regards cruise missile submarines will undoubtedly be the projected construction of a new type armed with the SS-NX-24 land-attack missile. The latter has a range currently estimated at 2,200 nautical miles, and is reported to be undergoing testing in Northern waters aboard a single converted unit of the 'Yankee' class. The SS-NX-24 is a much more substantial missile than the tube-launched SS-N-21 which has recently entered service on the 'Akula' and the 'Sierra' and carries a 1MT warhead. With a length estimated at 12.5 metres it requires a special launch tube. The trials unit has been extensively modified, and now has a large box-shaped missile compartment housing twelve fixed-angle launch tubes abaft the conning tower.

Since a purpose-built submarine designed to launch the SS-NX-24 missile would necessarily approach the size of the 'Oscar' itself, it may be that the new type, currently believed to be at an advanced stage of design, will replace the former on the building ways, as only the Severodvinsk Shipyard appears to be able to construct such large hulls at present.

As regards attack submarines, series production of the 'Sierra' and 'Akula' will probably continue at the Gorky and Komsomolsk Shipyards respectively. Komsomolsk is capable of turning out at least one 'Akula' per year, while production of the 'Sierra' at Gorky may be somewhat below this level. It is reported that only the hull of the first unit was built at the inland shipyard, and that fitting out subsequently took place at Severodvinsk, the incomplete hull being transported to the Northern Fleet area via the White Sea Canal.

There are reports that the 'Victor III' remains in production, although it is possible that the units currently being built are for export, as it thought that the 'Victor' is the obvious choice to meet India's requirement for a force of nuclear-powered submarines. The modified 'Charlie I' which has recently been supplied to the Indian Navy appears to have had its SS-N-7 missile tubes disabled, and is almost certainly for training. An order for six new-build 'Victors', probably with modifications to their sonar and fire control systems, is postulated.

If the export 'Victors' are being built at the Admiralty Shipyard, Leningrad – as seems likely, give that the other shipyard with experience of building the type, the Komsomolsk Shipyard, is now heavily committed to construction of the 'Akula' – this will undoubtedly delay any planned expansion in 'Sierra' production.

Production of the diesel-powered 'Kilo' continues at Komsomolsk at the rate of about two per year, the completed units having entered service with the Soviet Pacific Fleet. Production lines have now been opened in the western USSR at the Gorky and Sudomekh (United Admiralty) Shipyards, but the latter appears to be producing submarines exclusively for export at present. At the time of writing four 'Kilos' have been ordered by Poland, one (possibly two) by Romania, one by Algeria, five by India and one by Syria.

This constitutes a substantial production programme which will keep Soviet submarine building yards fully occupied for some considerable time. It will also necessarily reduce the supply of new-build submarines for the Soviet Navy in the near term, although this will in part be compensated by benefits accruing from the modernization of friendly and allied navies.

The success of the 'Kilo' as an export design serves to confirm its primary function as a replacement for older diesel-electric medium boats of the 'Whiskey' and 'Romeo' type. Several of the foreign navies that are now receiving the 'Kilo' have ordered these submarines specifically to replace older units belonging to one or other of these two classes.

Technological capabilities

The only major new weapon system expected over the next few years is the SS-NX-24 land-attack cruise missile, which is already undergoing trials aboard the converted 'Yankee'.

The Soviet Navy has now developed a particularly full range of offensive and defensive weaponry, and new missiles and torpedoes developed in the near term will almost certainly be modifications of existing types, with the customary incremental improvements.

Improvements should also be expected in passive acoustic technology, with considerable emphasis given to more advanced data processing techniques. The research work of the Soviet scientific institutions will undoubtedly be boosted, as in the past, by the fruits of

Above: *The 'Kilo' is proving to be a popular export design, and orders from friendly countries currently exceed the numbers being built for the Soviet Navy. This is the Indian* Sindhugosh, *delivered in 1986. (US Navy, 1986)*

industrial espionage and any other form of technology transfer. Despite frantic efforts to keep abreast of developments in the West, however, this is still an acknowledged area of Soviet weakness.

Soviet submarines will undoubtedly become quieter, to the point where Western navies will have to reconsider their dependence on passive sensors for anti-submarine warfare. This development will be a natural consequence of a fundamental change in the primary mission of Soviet attack submarines, for which the anti-submarine 'hunter-killer' mission is becoming as important as the more traditional anti-surface mission.

Improvements in the quieting of Soviet submarines will not, however, take place overnight. In US congressional testimony in 1985, Rear-Admiral John Butts, former Director of US Naval Intelligence, stated only that 'the Soviets have steadily *reduced the technological inferiority* [author's italics] of their latest submarines'. Vice-Admiral Bruce DeMars, Deputy Chief of Naval Operations for Submarine Warfare in the US Navy, has talked of the Soviets starting 'twenty years behind', and has noted the difficulty of accommodating advances in older boats on a piecemeal basis.[1]

Superior propeller technology has been identified as the one improvement which could be extended to existing submarines, and this is an area in which the Soviets are known to have had some success in acquiring Western technology. Other areas, however, such as the design and installation of propulsion and auxiliary machinery, will undoubtedly occupy years of research and experimentation.

The one area in which the Soviets appear to be well advanced is that of command, control and communications, which have always been accorded a high degree of priority in Soviet tactical doctrine. An extensive network of military satellites performing a variety of functions, from the monitoring of surface traffic to submarine navigation and communications, is maintained in place above the world's oceans. Nuclear power appears to be increasingly common as a means of extending the operational life of these satellites, and they are replaced at regular intervals as part of a large-scale launch programme.

The Soviets are also thought to be well-advanced in their programme to provide secure ELF communications for missile submarines of the 'Typhoon' and 'Oscar' classes. Three shore-based stations, each of which is more powerful than the single US Navy facility in Wisconsin, are reported to have been completed since 1983, of which one is thought to be located in the area of the newly constructed 'Typhoon' base at Gremikha, at the eastern end of the Kola Peninsula.

Conclusion

The submarine force has been the major offensive arm of the Soviet Navy ever since its inception, and will undoubtedly remain so for the foreseeable future. If anything, its importance has increased with the appointment of Admiral Chernavin.

The surface fleet so painstakingly built up by Gorshkov will continue to occupy an important place in Soviet maritime strategy, but it is difficult to envisage how it can maintain its recently acquired prestige in the face of increasing restrictions placed on out-of-area operations, and the imposition of a strictly defensive strategy in which surface forces will be strictly subordinated to land-based control employing Combined Arms tactics.

In a new era of *détente*, and with a leadership dedicated to improving the lives of ordinary people, the armed services of the Soviet Union can no longer rely on unlimited budgets for research, development and equipment. When the 'crunch' comes, and military programmes are subjected to a fundamental reappraisal, it may be that the Soviet surface fleet will have to bear the brunt of any cuts in order to fund the submarine programme. The history of the Soviet Navy since the Revolution of 1917 suggests that any other solution would be surprising.

References

Introduction

1. Admiral S. Gorshkov. *The Seapower of the State.* Naval Institute Press, Annapolis, 1979.
2. Vice-Admiral K. Stalbo. 'Some Issues of the Theory of the Development and Employment of the Navy', in *Morskoy Sbornik*, April/May 1981.

The Post-War Era

1. Jurg Meister. *Soviet Warships of the Second World War.* London, 1977.
2. Dr Robert S. Suggs. 'Soviet Subs in Scandinavia: 1930 to 1945', in USNI *Proceedings*, March 1986.
3. Siegfried Breyer. *Guide to the Soviet Navy*, 2nd ed., p. 282. Naval Institute Press, Annapolis, 1977.
4. Figures taken from Norman Friedman. *All the World's Fighting Ships, 1947–82.* Conway Maritime Press, London, 1983.
5. Figures quoted by Norman Polmar. *Guide to the Soviet Navy*, 4th ed., p. 93, Naval Institute Press, Annapolis, 1986.
6. Roy Corlett. 'Advances in Soviet Underwater Capability and Weapons', in *Maritime Defence*, June 1987.

The Nuclear Age

1. Figures taken from Michael McцGwire, quoted by Siegfried Breyer in *Guide to the Soviet Navy*, 2nd ed., p. 32, Naval Institute Press, Annapolis, 1977.
2. John G. Engelhardt. 'Soviet Sub Design Philosophy', in USNI *Proceedings*, October 1987.
3. Roy Corlett, op. cit.
4. ibid.
5. Siegfried Breyer. *Guide to the Soviet Navy*, 1st ed., pp. 52–3, Naval Institute Press, Annapolis, 1970.
6. Norman Polmar. *Guide to the Soviet Navy*, 4th ed., p. 126, Naval Institute Press, Annapolis, 1986.
7. ibid., p. 96.

The Anti-carrier Strategy

1. A projected total of 72 has been suggested by Michael MccGwire, quoted by Siegfried Breyer in *Guide to the Soviet Navy*, 2nd ed., p. 36, Naval Institute Press, Annapolis, 1977.

The New Generation

1. Norman Friedman. *Submarine Design and Development*, p. 105, Conway Maritime Press, London, 1984.
2. *Aviation Week and Space Technology*, 1977.
3. Roy Corlett, op. cit.

'Alfa' and 'Papa'

1. Norman Friedman. 'The Soviet Fleet in Transition', in USNI *Proceedings*, May 1983.
2. For a more detailed analysis *see* Käre M. Heggstad. 'Submarine Hulls of Titanium', in *Maritime Defence*, April 1983.
3. Norman Friedman. *Submarine Design and Development*, p. 134, Conway Maritime Press, London, 1984.
4. ibid.

The Bastion Concept

1. Roy Corlett. 'Advances in Soviet Underwater Capability and Weapons', in *Maritime Defence*, June 1987.
2. *Understanding Soviet Naval Developments*, 5th ed., p. 25, Office of the Chief of Naval Operations, US Navy, 1985.

The Missile Giants

1. Dan Boyle. 'ELF – An Essential Communications System', in *International Defence Review*, vol. 11, No. 7, 1978.
2. 'The Navy's Strategic Communications Systems – Need for Managerial Attention and Decision making', US General Accounting Office, Washington, DC, 1979.
3. Rear-Admiral E. Eikanger, Chief of Norwegian Intelligence, reported in *Jane's Defence Weekly*, 6 February 1988.

The New Attack Boats

1. *Defence*, p. 309, June 1987.

Postscript

1. Interview with Vice Admiral Bruce DeMars, US Navy, in USNI *Proceedings*, October 1987.